Exploring World Regions

Gerald Ingalls   Walter Martin

Associate Professor of Geography   Instructor in Geography

The University of North Carolina at Charlotte

**KENDALL/HUNT
PUBLISHING COMPANY**
Dubuque, Iowa

PREFACE

I hear and I am aware.
I see and I know.
I do and I understand.

As the proverb suggests, the learning process has several dimensions. Unfortunately, "doing" is a critical dimension that can all too frequently be overlooked in the introductory college level class. The "hands-on" effort is frequently difficult, if not impossible, to obtain in an introductory world regional geography class, where large enrollments, short class periods and voluminous amounts of subject matter provide ample justification for a lecture and test format. For this reason, we offer this combination of exercises, study guides, maps and data as an aid to those making the effort to appreciate the geographic variations in cultural, demographic, economic and political patterns.

We hope that the exercises promote better learning through doing. We hope that the regional study guides, the maps and the data facilitate the lecture and discussion that occurs in the classroom. We hope that the book makes taking and teaching the course a more pleasant and rewarding experience.

We wish to acknowledge the yeoman efforts of Connie Cook, Denise Askland, Jackie Johnson, Jeff Simpson, Miles Champion, Carl Flick, Tran Van Ra, Gary Addington, and Diane Bizzell in putting together the last drafts of this document. Their efforts were superb. We also wish to commend the efforts of Carol Rhea in data collection, proofing, and editing. To all we say thanks.

G.L. Ingalls
W.E. Martin
June, 1983

TABLE OF CONTENTS

INTRODUCTION

We are confronted daily with a complex array of cultures, lifestyles, problems and conflicts from any and every point on the globe. Televisions, radios, newspapers, magazines, books and telephones constantly bombard us with the visual, verbal and written images of places near and far. We are confronted not only with conflict but also with more global, and perhaps ultimately even more serious challenges such as environmental pollution, depletion of vital resources, too many people for too few resources, ignorance and intolerance of social, ethnic racial, economic and cultural differences.

Each of us has our own system for dealing with the flood of complex, often overwhelming, informtaion with which we are faced daily. Some of us choose to retreat into the security of narrowly defined dogma which afford us comfortable, convenient and often simplistic organizational schemes for coping with the daily deluge. Unfortunately, the black/white, either/or answers which are the hallmark of such dogmatism can, more often than not, fall short of adequate explanation.

An alternative means of coping with the voluminous flow of world events is to adopt a broader perspective which involves searching out an organizational scheme for compartmentalizing, managing and digesting the information that comes our way. Under this broader, more global perspective, we learn to appreciate the values, points of view and ways of life of other cultures and places. Such a perspective comes by reading, watching and listening to a wide range of divergent viewpoints over many topic areas. The value of this approach is that it broadens our horizons and promotes a more flexible, less dogmatic, means of dealing with the complex problems and uncertainties of today's world.

In reaching for this global perspective geography, and even more specifically, regional geography, has something critical to offer. As one of the oldest and most durable themes of geographic study, regional geography affords the type of organizational scheme so vital in coping with the daily flow of worldwide, diverse information. As such, many colleges and universities offer an introductory course which surveys a number of regions of the world focusing on both the common and unique conditions of each.

We have written this book with such a course in mind. In the belief that learning does not stop or start as students walk out or into the classroom, we have structured an aid to study outside of the formal classroom setting. We have structured a book that has a series of exercises, study guides, maps and data which serve to complement the standard world regional text.

Exploring World Regions is designed as a supplement to most standard introductory texts for World Regional Geography. *Section One* and *Section Two* survey three of the world's most difficult and pervasive problems: population growth, food supply, and economic development. We believe these problems are fundamental to understanding the international mosaic of landscapes and regions. Beginning with the Demographic Transition model and ending with a more detailed assessment of these regional disparities, seven exercises are designed to lead the reader toward grouping the nations of the world into regions with relatively homogeneous demographic and economic characteristics. *Section Three*, employs a conventional classification of contiguous world regions to organize study guides for nine of the most populous regions. Each of these study guides include a list of basic concepts, a list of place names, and a mapping activity. These nine regions and their nations were selected because they illustrate the problems and characteristics of modernized places, traditional places, and those along the continuum between those two worlds. The Atlas Section, *Section Four*, and the Characteristics of Nations, *Section Five*, are included to provide a convenient up-to-date reference for locating and comparing places.

It is our hope that through knowing and doing you will understand and that by understanding the nature of places, you will be curious, more interested, and more perceptive about our situation in the world. It is our hope that the explanations you achieve from this approach will make you more comfortable in dealing with the deluge of information with which you are faced daily.

Table 1

Quick Reference Index to Maps and Countries

Country	Map Number	Country	Map Number
Afganistan	11	Japan	5
Albania	3	Jordan	11
Algeria	11	Kuril Islands	5
Andorra	3	Kuwait	13
Argentina	15	Laos	6
Anguilla	17	Lebanon	11
Antigua	17	Leichtenstein	3
Aruba	17	Libya	11
Australia	9	Luxembourg	3
Austria	3	Malaysia	6
Bahamas (The)	17	Malta	3
Bahrain	13	Martinique	17
Balearic Islands	3	Mexico	16
Barbados	17	Monaco	3
Belgium	3	Mongolia	9
Belize	16	Morocco	11
Bolivia	15	Netherlands	3
Bonaire	17	Nevis	17
Brazil	15	Nicaragua	16
Bulgaria	3	North Korea	5
Burma	6	North Yemen	11
Cayman Islands	17	Norway	3
Chile	15	Oman	11
China	9	Pakistan	11
Colombia	15	Panama	16
Corsica	3	Paraguay	15
Costa Rica	16	Peru	15
Crete	3	Philippines	6
Cuba	17	Poland	3
Curacao	17	Portugal	3
Cyprus	11	Puerto Rico	17
Czechoslovakia	3	Qatar	13
Denmark	3	Romania	3
Dominica	17	St. Christopher	17
Dominican Republic	17	St. Lucia	17
Egypt	11	St. Vincent	17
El Salvador	16	Sardinia	3
Equador	15	Saudi Arabia	11
Falkland Islands	15	Sicily	3
Faeroe Island	3	Singapore	6
Federal Republic of Germany	3	South Korea	5
Finland	3	South Yemen	11
France	3	Spain	3
French Guiana	15	Sudan	11
German Democratic Republic	3	Suriname	15
Greece	3	Sweden	3
Greenland	3	Switzerland	3
Grenada	17	Syria	11
Guadelupe	17	Taiwan	6
Guatemala	16	Thailand	6
Guyana	15	Trinidad and Tobago	17
Haiti	17	Tunisia	11
Honduras	16	Turkey	11
Hungary	3	Turks and Caicos Islands	17
Iceland	3	U.S.S.R.	8
India	9	United Arab Emirates	13
Indonesia	6	United Kingdom	3
Iran	11	United States	2
Iraq	11	Uruguay	15
Ireland	3	Venezuela	15
Israel	11	Vietnam	6
Italy	3	Virgin Islands	17
Jamaica	17	Yugoslavia	3

Section 1
Population and Food Supply

INTRODUCTION

Until some 200 years ago the size of the human population remained fairly stable because high birth rates were balanced by high death rates. The great demographic transition came when death rates fell.

Ansley J. Coale, 1974

In two brief sentences, Ansley Coale summarizes the essential elements of one of the most serious problems facing humankind today. While technological advances in medicine have increased life expectancy, the majority of the world's billions of people still are convinced that two kids are not enough! For economic, cultural or personal reasons, larger families are still viewed as necessary, or desirable, or both; hence, while death rates have declined markedly in recent decades, there has been, for a vast majority of the world's population, no corresponding incentive to lower birth rates. The overall rate of natural increase of the world's population has soared—so much so that labels like explosion, crisis and bomb are inextricably tied to most current assessments of worldwide population growth. But, as Coale implies, it has not always been so. Given the one million or more years of human existence, the last two centuries of rapid population growth is but a "temporary deviation" (Freedman and

Berelson, 1974).

For most of its history, humankind reproduced slowly—virtually at zero levels of growth. Primitive levels of technology and a hunting and gathering system imposed natural limits on population growth. However, as is indicated in Table 2, the coming of the Agricultural Revolution, usually dated at 8000 B.C., altered dramatically the relative balance between birth and death rates.

As Weeks (1977) argues, it was not so much that the settled lifestyle of stable agricultural and urban communities fostered lower death rates through improved living conditions. Rather, anthropological evidence suggests that, as the Agricultural Revolution began, death rates probably also increased as humans settled into higher density settlements with attendant sanitary problems and increased levels of exposure to diseases. Death rates increased but birth rates increased more. Improved diets increased the ability and the opportunity of women to conceive and to bear children successfully. Regardless of the root causes, rates of natural increase began, at least initially, a slow but steady climb upward.

While the Agricultural Revolution chipped away at barriers to rapid population growth, the Industrial Revolution demolished them almost completely. As recently as 1750, the world's population was growing at the relatively slow annual rate of 0.05 percent. At that rate, it would have required over 1200 years to double the estimated 800 million population. But the material and intellectual benefits of the Industrial Revolution, which was in its early stages by 1750, drastically altered the state of events. A rising standard of living, with more food, better living conditions, particularly in personal hygiene, and sanitary matters acted in concert to lower death rates. The age of discovery and exploration opened new lands for agricultural exploitation and for a seemingly endless source of resources to feed the industrial fires. And as populations soared in the industrializing European cultural areas, these new lands acted to syphon off excess millions in a wave of migrations. Innovations in transportation, energy use and in power-driven machinery fostered communication and movement to tie new worlds to old and make movement of people and ideas increasingly easier. By the late 19th and early 20th century, innovations in science had accelerated the desire to attack all forms of human limitations, including disease. The resulting advances in medical technology, particularly in this century, dramatically lowered death rates. Birth rates remained, for all but

industrial societies, comparatively stable and high.

By 1800 A.D. the world had its first one billion population, a goal which had required over one million years to obtain. Not so the second billion. That goal was reached a short 127 years later; 33 years later, the third billion was in place; the fourth billion required but 15 years; and at current rates of growth, the fifth and sixth billions will be in place by the year 2000!

The dimensions of the problem are clearly outlined in statistics such as those presented in Table 2. The earth's population is doubling in size over ever-decreasing spans of time. At current rates of growth, there are 231,280 more people each day; approximately every 34 days enough people are added to the world's population to match the total increase of the first one million years (to 8000 B.C.) of human existence. Every year there are the equivalent of 10 new cities the size of Moscow.

It is not so much the shear number of people being added to the earth's population, although there are many who agree with the argument posed by Lester Brown (1978) that the pressure on the earth's biological systems are impairing its reproductive capacities. Since humankind first settled into agricultural communities, the potential for food production has increased steadily and markedly. As Brown and Finsterbusch (1971, p. 57) argue, "an adequate diet has been assured for most of that third of mankind living in North America, Western Europe, Eastern Europe, and Japan." Herein lies the crux of the problem. The most immediate problem posed by the world's rapidly expanding population is not numbers of people but their distribution with regards to food resources.

Unfortunately the people being added are not entering economic or cultural systems which are as capable of providing food and resources to their populations as are the systems of which Moscow and New York City are a part. As is depicted by the data summarized in Table 3, the pattern of population growth is not uniform. Among the 10 nations with the largest numbers of people, the rate of population growth varies from a low of 0.7 percent to a high of 3.3 percent. The U.S. and the U.S.S.R. are among the world's most populous countries but have relatively low rates of growth. China has a rate of growth less than the world average, yet with a base of one billion people, even the relatively moderate rate of increase of 1.4 percent represents 15,349,500 new inhabitants each year. That is the equivalent of adding the population of two New York City's each year! Such an increase would strain even the most

8

Table 2
Growth of World Population

Date	Total Population (in millions)	Growth Rate (% Increase)	Years Required to Double Population From One Date to the Following Date
8000 B.C.*	8	.0015	—
A.D. 1*	300	.0360	1530
1750*	800	.0560	1240
1800*	1000	.4400	—
1927**	2000	.6800	102
1960**	3000	1.2300	57
1975**	4000	1.9200	36
1982***	4667	1.8000	39
2000***	6130	—	—

Sources:

* Ansley J. Coale, "The History of the Human Population," *Scientific American*, Vol. 231, No. 3, September 1974, 41-51.

** William W. Murdoch, *Environment, Resources, Pollution, and Society*, (Sunderland, Mass., Sinauer Associates, 1975), p. 49, as quoted in R. Jackson and L. Hudman, *World Regional Geography*, John Wiley, New York, N.Y., 1982, p. 71.

*** *1983 World Population Data Sheet*, Population Reference Bureau, Inc., Washington, D.C., 1983.

Table 3
The World's Ten Most Populous Countries*

Rank	Country	Population Estimate (in millions)	Rate of Growth (% per Year)	Years to Double
1	People's Republic of China	1,023.3	1.5	46
2	India	730.0	2.1	33
3	U.S.S.R.	272.0	0.8	83
4	U.S.	234.2	0.7	95
5	Indonesia	155.6	1.7	41
6	Brazil	131.3	2.3	30
7	Bangladesh	96.5	3.1	22
8	Pakistan	95.7	2.8	25
9	Nigeria	84.2	3.3	21
10	Mexico	75.7	2.6	27

*Source: *1983 World Population Data Sheet*. Population Reference Bureau, Inc., Washington, D.C., 1983.

productive and efficient agricultural system.

Why is there such a difference in the rates of population increase for these nations? Are there distinctive patterns that can be observed in the variation of rates of natural increase of the population of nations? If so, what can we learn from these patterns? And what if we examined the distribution of food production? Are there areas of surplus and deficit? What relationship do the food production patterns have to the patterns of population in growth?

Such questions are the natural focus of one aspect of geographic inquiry. Variations in any phenomena are what geographers seek to explain. In the next three exercises you will be asked to complete some elementary forms of pattern analysis to account for variations in distributions of several phenomena. The three exercises will deal with variations in the rates of population growth and in the production of food resources needed to sustain population growth.

The first of these exercises deals with one model used to assess variation in birth rates and death rates across time and across space. As it is used in Exercise 1, the Demographic Transition Sequence is a means of examining spatial variation at one point in time. It is a means of comparing nations and regions according to their rates of natural increase in population.

The second of the exercises in this section deals with the central question of population growth—what is the ability of the earth's environment to support the population that is here now and those people who are yet to come? As is the case with population growth rates, the central issue is spatial variation; however, in this case, you are asked to examine variations in the patterns of food production.

Not all nations are equally capable of producing sufficient food to feed their population. Not all economic and cultural systems are equally capable of full and complete utilization of their physical environments. And for some economic and cultural systems, even full and efficient utilization may prove insufficient as the shear weight of massive populations overtaxes already overburdened environmental systems. Consider China, India and Bangladesh as cases in point. In Exercise 2, you are asked to explore the relationship between food production and population growth in a sample of nation states.

Finally, in Exercise 3, the international patterns of food production and consumption for selected grains, meats and foodstuffs are examined. The patterns of surplus and deficit and the flow to equalize or balance the system are scrutinized.

EXERCISE 1

The Demographic Transition Sequence:
A Model of Population Growth

What began as an effort to describe the changes in birth and death rates and overall rates of natural increase in economically advanced nations has gradually evolved into a model utilized to account for the demographic changes wrought by economic development. Since the earliest efforts of Warren Thompson in 1929, demographers have observed, measured and detailed the types and sequence of changes in birth rates that accompany the development of a nation from a traditional, or lesser developed society, to a modern, industrial or post-industrial society. From these efforts has emerged a sequence of four stages or categories marked by distinct levels of birth and death rates.

In Stage 1, birth rates and death rates are both high. In the period before 1750 A.D., most of the world fit the characteristics of this stage. Birth rates and death rates exceeded 30 per 1000 and consequently largely offset one another. The end result was slow (or virtually no) growth. As a consequence, populations remained relatively small

and more susceptible to natural events, such as environmental catastrophes, and human events, such as war and disease.

In Stage 2, birth rates remain high; rates of 30 or more per 1000 are not at all uncommon. However, as medical technology and personal and public hygiene improve, the death rate falls off markedly. In developing many of the earlier classic demographic transition models (based on European industrial societies), demographers such as Thompson hypothesized that death rates fell because of improved hygiene, more abundant food supply and better living conditions. In most of the traditional, or third world societies, which have recently entered Stage 2, improved medical technology—especially since 1930—has increased overall longevity and reduced infant mortality to the point that death rates have dropped significantly. However, the incentives, especially economic incentives such as those that accompany industrialization with its associated urbanization, have not accompanied the lower death

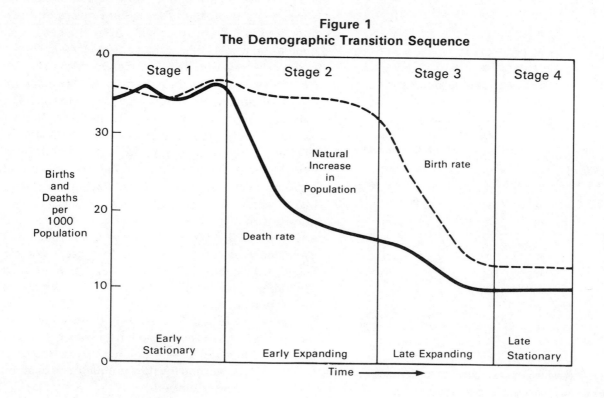

Figure 1
The Demographic Transition Sequence

10

rates. As a consequence, birth rates remain quite high at 30 to 40 or more per 1000. In Figure 1, the demographic transition sequence is graphically depicted. As is evident, the lines representing birth and death rates grow further apart as death rates fall and birth rates do not, during Stage 2. The further apart the birth and death rate lines, the higher the rate of natural increase.

During Stage 3, the birth rate begins to decline; the death rate also declines somewhat from the levels obtained in Stage 2, and gradually levels off at a low plane. Again under the classic European models used to formulate the original transition sequence, declining birth rates accompanied industrial and urban lifestyles. In urban, industrial, and economically developed societies, the need for large families decreases as society provides for old age security, lowers infant mortality, and removes children from the labor force by instituting child labor laws and compulsory education. Children become economic liabilities rather than the assets they are in a more traditional agricultural setting. The birth rate falls markedly during this stage and approaches, but does not match, the low levels of the death rate. However, the overall rate of increase remains high during most of Stage 3, since the difference (distance between the lines on the graph in Figure1) between birth and death rates remains relatively great during most of this stage of the transition.

In the final stage of the transition sequence, as the social, economic and technological processes begun in Stage 3 continue, birth and death rates stabilize at a fairly low level. This stage is one that has been achieved by only the most mature urban-industrial societies. Among Stage 4 societies, birth rates average from about 10 to 20 per 1000 and death rates average around 10 to 15 per 1000.

The demographic process outlined above describes population changes that accompany economic changes. And, as originally proposed, the transition sequence was intended to depict the changes that occur in demographic structure during economic development. However, the demographic transition model is also useful in examination of the differences that exist in demographic structure in a cross-sectional or regional sense—in other words, to freeze time and examine differences in stages of demographic transition across space. In this exercise, you are asked to compare the differences in fertility, as measured by crude live birth rate, and mortality, as measured by crude death rate. The difference between these two statistics, as measured by the rate of natural increase, or the excess of births over deaths, is the basic element of the demographic transition sequence.

In this exercise, you are asked to utilize actual data on the 1983 birth and death rates of selected nation states around the world in order to determine if there are nations which represent all four stages of the demographic transition sequence. Table 4 contains vital statistics for selected countries in 1983. On the world outline map provided, you are to graphically represent the spatial pattern of demographic transition for the nations represented in Table 4. Classify each nation as Stage 1, 2, 3, or 4. Use the following rates of birth and death to categorize nation states:

	Birth Rate	Death Rate
Stage 1	35 or higher	32 or higher
Stage 2	27 or higher	32 or less
Stage 3	20 to 27	15 or less
Stage 4	20 or less	15 or less

Color or shade within the outline of each nation according to whether it is Stage 1, 2, 3, or 4. Pick four distinctive (but preferably blending) colors and use one to represent each stage of transition. After you have completed your map, examine the resulting patterns and respond to the questions on the next page. Use the population and economic indicators in Section 5 to assist you in answering these questions.

Table 4

Birth and Death Rates for Selected Countries*

Country	Birth Rate	Death Rate	Country	Birth Rate	Death Rate
Egypt	37	10	Japan	13	6
Spain	13	7	Korea, South	23	8
Ghana	47	15	Taiwan (1)	21	5
Mali	49	21	Canada	15	7
Nigeria (2)	48	17	United States	16	9
Upper Volta	48	22	Mexico (2)	32	6
Ethiopia	43	22	Haiti	36	13
Angola	47	22	Bolivia	42	16
South Africa	35	14	Brazil	31	8
Israel	24	6	Venezuela (2)	33	6
Saudi Arabia (2)	42	12	Argentina	24	8
Turkey	35	10	Denmark	10	11
Yemen, North	48	21	Sweden	11	11
Bangladesh	45	17	Germany, West	10	11
India	34	13	Germany, East	14	13
Nepal	42	18	Romania	15	10
Burma	37	15	Italy	11	10
Singapore (1)	16	5	Yugoslavia	17	10
China (PRC)	19	8	Australia	16	7
Hong Kong (1)	15	5	U.S.S.R.	20	10

*Source: *1985 World Population Data Sheet,* Population Reference Bureau, Inc., Washington, D.C., 1985.

EXERCISE 1
Answer Sheet

1. Are there any Stage 1 nations? If so, list each. If not, explain why there are none.

2. Describe the pattern in the location of Stage 2 nations.

    What are some of the locational characteristics you can observe that Stage 2 nations have in common (continents, latitude, longitude, landlocked positions, etc.)?

    What is the significance of each of these locational attributes you have listed?

    Using the data in Section 5, do you see other social, economic or demographic indicators that Stage 2 nations have in common?

3. What do Stage 3 nations have in common? Is there a geographic pattern that is evident in the location of Stage 3 nations? If so describe it. If not, use the data in Section 5 to describe what other characteristics Stage 3 nations have in common.

4. In most introductory texts, there is a scheme for classifying nations according to stages of economic development. In this workbook, Section 2 will cover some methods of measuring economic development. These schemes are usually some form of a continuum ranging from developing (third-world, traditional, etc.,) to developed (modern,industrial, first- or second-world, etc.) societies. Most classification schemes do not provide a niche for the newly industrialized societies (such as Singapore) and the resource wealthy societies (oil-exporting countries such as Kuwait). In Table 4, these societies are identified by a 1, for NIC's and a 2, for oil exporting.

   Where are the NIC's on the demographic transition sequence? Why?

5. Where are the oil-exporting states on the demographic transition sequence? Why?

6. List the 4 nations which are not NIC. What social and economic characteristics do these nations have in common? Do China's social and economic characteristics match the others you have listed in stage 4? Explain the differences.

14

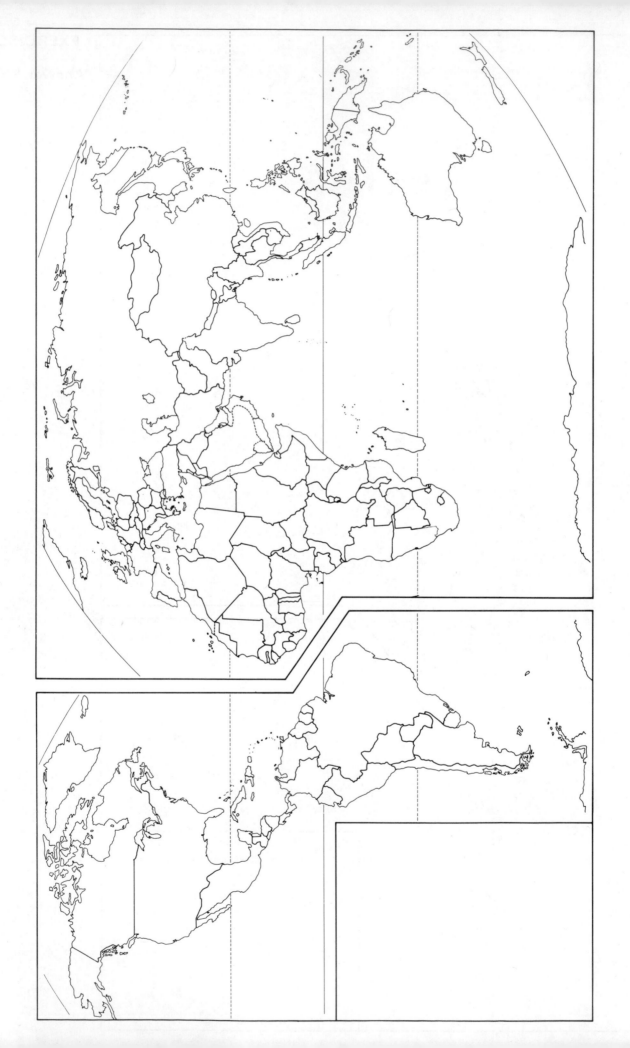

EXERCISE 2

The Growth of Food and Population

The earth's grasslands are a rich source of protein, from which comes most of the world's meat, butter, and cheese . . . they sustain the draft animals that till a third of the world's croplands . . . They supply food, industrial raw materials . . . and a variety of fibers, alcohols, starches and vegetable oils . . . Four billion human beings with rising expectations exert great pressure on these biological systems, often exceeding nature's long-term carrying capacity . . .

In some places, particularly in relatively poor countries, population growth now acts as a double-edged sword, expanding the demand for a given biological resource even as it reduces production.

Lester R. Brown, 1978

In the opening pages of his book, *The Twenty Ninth Day,* Lester Brown summarizes succinctly the exact dimensions of the population "problem." It is not simply a question of too many people. It *is* a question of too many people exerting too much pressure on environmental systems which are often incapable of bearing the weight of such sustained pressure. In purely Malthusian terms, population growth is outstripping food supply in some areas of the globe. Now that you have completed Exercise 1, perhaps you already have a reasonable good idea of where Malthusian predictions are in danger of coming true.

In another publication, Lester Brown (1983) argues that a combination of declining oil production, deteriorating biological systems, and shrinking per capita cropland acreage is slowing economic growth and is changing the nature of the population problem. At this point, rapid population growth is inhibiting economic expansion in some countries and thereby precluding any improvements in income. The result is that, for many countries, living standards are falling as economic conditions deteriorate. The countries caught in this dilemma are essentially frozen on the demographic transition and economic sequence. Death rates have fallen, but unfortunately social and economic conditions still entice people to have larger families and birth rates remain high.

In this exercise, you are asked to explore the dilemma of a set of nations caught in this demographic and economic "squeeze." You are asked to explore the relationship between food supply and population growth rates. This exercise is designed specifically to help you focus on some of the difficulties facing impoverished nations with rapidly-expanding populations.

Another aspect of this exercise is the technique used to examine the relationship between two variables. In situations such as this, where questions of the relationship between two phenomena are posed, it is often instructive to directly compare how the phenomena vary. This can be accomplished in many ways. For example, you could construct a second map of food (total) production and compare any resulting patterns to your map depicting countries at various stages of the demographic sequence. This procedure is, unfortunately, rather subjective when any evaluation or assessment is necessary.

In this exercise, you are asked to utilize another means of assessing the covariation or relationship between two variables. You are asked to plot *graphically* the variation in two variables. The two variables which you are to compare are the food index and the rate of natural increase in the population of selected nation states.

The food index is a quantitative measure of the food produced in 1980, compared with the amount of food produced in that country 10 years earlier. The index describes food production per capita. A score of 100 on the food index means that exactly as much food was produced in that nation in 1980 as in 1970 on a per capita basis. Scores of more than 100 indicate the nation has made a net gain over the last decade while scores of less than 100 indicate that less food was being produced in 1980 than a decade earlier. For example, Mozambique, with a food index of 75 produced only 75% (3/4) as much food per capita as it did in 1970. The food index definition of food includes cereals, starchy roots, sugar cane, sugar beets, pulses, edible oils, nuts, fruits, vegetables, livestock and livestock products, but not animal feed, seed requirements, or food lost in processing or distribution. The food index describes the relative growth or decline in domestic food production.

The rate of natural increase (R.N.I.) is calculated by subtracting the crude death rate (CDR) from the crude birth rate (CBR) and dividing by 10. The rate of natural increase is the percentage by which the population will grow in one year discounting migration.

In Table 5, there are 40 nations listed. These are the same 40 nations listed in Table 4 of Exercise 1. On the graph provided on the next page, the axes are labeled Food Index and Rate of Natural Increase. Use this graph to plot the position of each of these 40 nations with regard to these two variables. After you have plotted each nation, use your graph to respond to the questions on the following page.

Table 5

Rates of Population Growth Compared to Increases in Food Production

Country	Natural Increase*	Food Index**	Country	Natural Increase*	Food Index**
Egypt	2.7	90	Japan	0.6	91
Spain	0.6	125	Korea, South	1.7	134
Ghana	3.2	74	Taiwan	1.6	-NA-
Mali	2.8	88	Canada	0.8	109
Nigeria	3.1	91	United States	0.7	116
Upper Volta	2.6	94	Mexico	2.6	106
Ethiopia	2.1	85	Haiti	2.3	89
Angola	2.5	81	Bolivia	2.7	102
South Africa	2.1	104	Brazil	2.3	125
Israel	1.8	103	Venezuela	2.7	104
Saudi Arabia***	3.0	69	Argentina	1.6	116
Turkey	2.5	112	Denmark	-0.1	111
Yemen, North	2.7	92	Sweden	0.0	117
Bangladesh	2.8	96	Germany, West	-0.2	119
India	2.2	103	Germany, East	0.1	110
Nepal	2.4	84	Romania	0.5	147
Burma	2.2	102	Italy	0.1	112
Singapore	1.1	148	Yugoslavia	0.7	117
China (PRC)	1.1	116	Australia	0.9	117
Hong Kong	1.0	71	U.S.S.R.	1.0	102

*Source: *1985 World Population Data Sheet,* Population Reference Bureau, Inc., Washington, D.C., 1985.

**Source: *World Development Report, 1983,* The World Bank, London, Oxford University Press, 1983.

***Data for 1981

-NA- Data Not Available.

Answer Sheet

1. Which 10 nations have seen the largest per capita gains in food production over the last decade? In what state of the transition sequence are these nations? Are these rich or poor nations (see per capita GNP, in Section 5)?

2. Which 2 nations have lost the most ground in per capita domestic food production? Are these poor nations? Do you think these nations are producing as much food in absolute terms as 10 years ago? Why or why not? Compare these two nations to Mali. All three nations lost ground. Why is Mali different from the other two?

3. Did the People's Republic of China have more domestic food production in 1980 than in 1970? Did India? Compare the two countries. Why are they different?

4. List the nations with low rates of natural increase (under 1.0) and declining food production per capita. What do they have in common?

5. List the nations with high rates of natural increase (over 2.5) and declining food production per capita. What do they have in common?

6. List the nations with high rates of natural increase (over 2.5) and increasing food production per capita. What do they have in common?

7. Do the points seem to be aligned in a general linear pattern? Draw in a line which approximates this general linear pattern. What is the significance of this line? Based on this line and your general observations, describe the relationship between food production per capita and rates of natural increase.

8. According to the food index, which of the world's ten most populous nations listed in Table 3 of Exercise 1, are expanding their agricultural productivity?

Answer Sheet

Graph Work Sheet

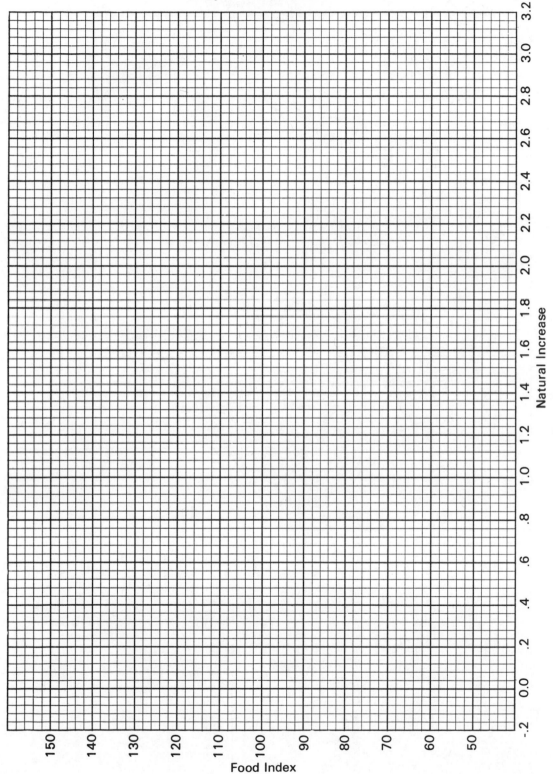

Food Index

Natural Increase

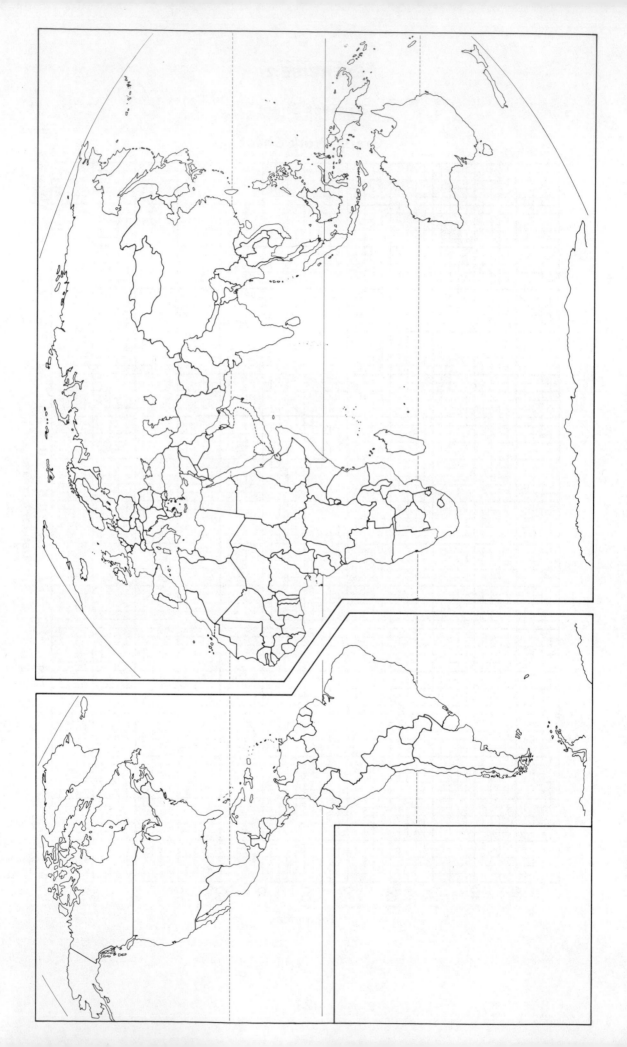

EXERCISE 3

The Availability of Food: Worldwide Patterns of Production and Consumption

Man's existence in relation to his food supply has always been precarious. But largely as a result of United States' initiative, the world has been spared from massive famine due to natural causes in recent decades.

L. Brown and G. Finsterbusch, 1971

One of the most fundamental and interesting aspects of world food production is the international pattern of food surpluses and imports. As Larry Brown and Gail Finsterbusch intimate, it is the tremendous surplus generated by "bread basket" regions in nations like the U.S., the U.S.S.R., Canada and Australia that sustain regions where deficits, compounded by too rapid population growth rates, threaten chaos. Exercise 1 indicated, for a selected sample of nations, differences in rates of population growth. Exercise 2 indicated, for the same selected sample of nations, that in many areas population growth outstrips food production. In this exercise, the focus is on the exchange that must take place in order to balance the ledger, i.e., to connect areas of surplus food supply, usually areas with relatively low population growth rates, to areas of deficit food supply—often areas of moderate to high population growth rates. This exercise is designed to emphasize the interdependence of nations and to acquaint you with the regional patterns of production, consumption, and the corresponding exchanges of food and feed. While other types of food comprise important shares of agricultural production, cereal grains, which are grasses with large seeds, are the most important single category (Table 6). This exercise emphasizes current patterns of wheat, rice, corn, and meat production and consumption. Wheat, corn, and rice represent 80 percent of the 1982 world grain harvest. Wheat is the single most important crop (28 percent of total grain production), with corn second (27 percent), and rice third (25 percent).

Wheat

Wheat is the most widely grown crop in the world. Wheat needs a growing season of at least 90 days and a minimum of nine inches of precipitation per year. Precipitation is beneficial during the early weeks of growth, but too much precipitation during the ripening period causes the spread of destructive diseases and insects. Consequently, two-thirds of all wheat acreage is located in relatively dry environments.

Table 7 provides basic data on world wheat production. Use this data to respond to questions 1—8 on Answer Sheet 1 of Exercise 3.

Table 6
World Agricultural Production

Commodity	1982 Share
Grains	30.4
Pulses	1.4
Roots and Tubers	5.8
Oilseeds	5.6
Vegetables	1.4
Fruits and Nuts	5.1
Meats	24.8
Milk	12.9
Eggs	3.8
Sugar	3.0
Fibers	3.2
Other*	2.8
Total	100.0

*Mostly tea, coffee, tobacco, and spices.

Source: U.S. Department of Agriculture, Economic Research Service, *World Agriculture: Outlook and Situation*. Washington, D.C.: U.S. Government Printing Office, March 1983.

Rice

As wheat is the preferred cereal grain of the wealthy technologically advanced nations, rice is the predominate cereal grain in many technologically developing nations. Rice requires a growing season of at least three months with a mean temperature of at least 75°F. Rice is much more tolerant of abundant rainfall and flooding than most crops and consequently is grown in the humid subtropics, tropical rainforest, and tropical savanna where abundant surface water supplies reduce cultivation

Table 7				Table 8		

Wheat: World Production and Consumption 1982/1983** | **Rice: World Production and Consumption 1982/1983****

Country	Pro-duction	Con-sumption	Country	Pro-duction	Con-sumption
Major Exporters	Million Tons		Major Exporters	Million Tons	
United States	76.4	23.8	United States	5.0	2.4
Canada	27.6	5.2	Thailand	11.4	8.6
Australia	8.5	4.0	Pakistan	3.1	2.1
EC-10*	59.4	44.8	China	102.0	101.4
Argentina	14.0	4.2	India	45.0	46.8
Major Importers			Burma	8.8	8.0
U.S.S.R.	86.0	105.5	Japan	9.4	10.7
China***	63.0	76.0	Italy	.6	.3
Eastern Europe	33.8	35.8	Australia	.4	.1
Other W. Europe	8.5	9.1	Major Importers		
Turkey	13.8	14.0	Indonesia	22.3	23.1
Brazil	1.8	6.1	South Korea	5.2	5.5
Mexico	4.2	4.2	Bangladesh	13.5	14.0
Other Latin Am.	1.4	7.9	Vietnam	9.0	9.0
Japan	.7	6.0	Other Asia	16.2	14.0
India	37.8	39.8	U.S.S.R.	1.6	16.9
South Korea	.1	2.1	Brazil	6.1	2.3
Indonesia	0	1.5	Other Latin Am.	4.7	6.5
Other Asia	16.4	22.9	Iran	.7	4.9
Egypt	2.0	7.9	Other N. Africa/ Mideast	1.9	3.7
Morocco	2.2	4.0	Malagassy	1.3	1.6
Other N. Africa/ Mideast	11.0	23.6	Nigeria	.9	1.5
Other Africa	3.4	7.0	Other Africa	1.8	3.6
Residual	.4	2.9	Residual	.6	2.1
World	472.4	458.3	World	271.5	276.4

* Belgium, Denmark, France, Greece, Ireland, Italy, Luxembourg, the Netherlands, United Kingdom, West Germany.

** Approximate data.

*** Since 1979, the Chinese government has encouraged farmers to grow and sell more produce in local markets. The result has been a 12% increase in the past seven years. Today China has surpassed the Soviet Union as the world's largest wheat producer.

Source: U.S. Department of Agriculture, Economic Research Service, *World Agriculture: Outlook and Situation.* Washington, D.C.: U.S. Government Printing Office, March 1983.

costs. Rice also thrives in desert and semi-arid lands where irrigation water is available.

Table 8 provides basic data on world rice production. Use this data to respond to questions 9-18, on Answer Sheet 2 for Exercise 3.

Coarse Grains

Coarse grains include corn, sorghum, barley, and oats. Corn is an important food staple in many countries. It dominates as the leading food energy crop in Central America, Andean South America, and East and South Africa. A large proportion of the corn harvest in Europe and North America is used as animal feed. Drought resistant grain sorghum has been increasingly accepted as feed for livestock and now ranks second only to corn as feed grain in the United States. Table 9 provides basic data on world production of coarse grains. Use the data from this table to respond to questions 19-22 on Answer Sheet 3 for Exercise 3.

Livestock

Second only to cereal grains in importance (Table 6), meat production represents approximately 25 percent of the world agricultural output. This production and consumption pattern strongly differentiates between the modernizing and the traditional world. With one-half of the world's population, the developing countries produce less than one-fifth of all livestock products. Table 10 shows the basic pattern of per capita meat consumption. Use the data from this table to answer questions 23-26 on Answer Sheet 4 for Exercise 3.

Table 9

Coarse Grains: World Production and Consumption

1982/1983*

Country	Pro- duction	Con- sumption
Major Exporters	Million Tons	
United States	255.5	157.3
Canada	26.6	18.5
Austrailia	4.2	2.9
Argentina	16.3	6.6
Thailand	3.7	1.7
South Africa	9.2	8.3
Major Importers		
U.S.S.R.	85.0	98.0
China	85.0	87.5
Eastern Europe	67.0	70.5
EC-10	70.4	74.9
Other W. Europe	21.3	31.9
Brazil	24.1	23.3
Mexico	11.3	20.8
Venezuela	1.3	2.7
Other Latin Am.	-2.1	8.0
Japan	.4	18.7
Taiwan	.1	4.1
South Korea	.9	4.4
Other Asia	42.1	44.5
Egypt	3.8	5.2
Iran	1.2	2.4
Israel	—	1.3
Other N. Africa/ Mideast	18.3	22.9
Other Africa	29.0	31.5
Residual	.6	-3.4
World	785.3	746.6

— negligible

* Approximate data

Source: U.S. Department of Agriculture, Economic Research Service, *World Agriculture: Outlook and Situation,* U.S. Government Printing Office, Washington, D.C., March, 1983.

Table 10

Summary of World Meat Consumption
(Killograms per capita per year)

Major Consumers	Beef and Veal	Pork	Poultry
United States	42	29	29
Canada	42	28	23
Mexico	18	17	7
Brazil	16	7	10
EC-10	24	34	14
U.S.S.R.	26	19	10
Japan	5.5	14	11
Other Nations			
China	2	15	3
Egypt	0.3	1	3
India	0.3	0.1	0.2

Source: U.S. Department of Agriculture, Economic Research Service, *World Agriculture: Outlook and Situation,* U.S. Government Printing Office, Washington, D.C., March, 1983.

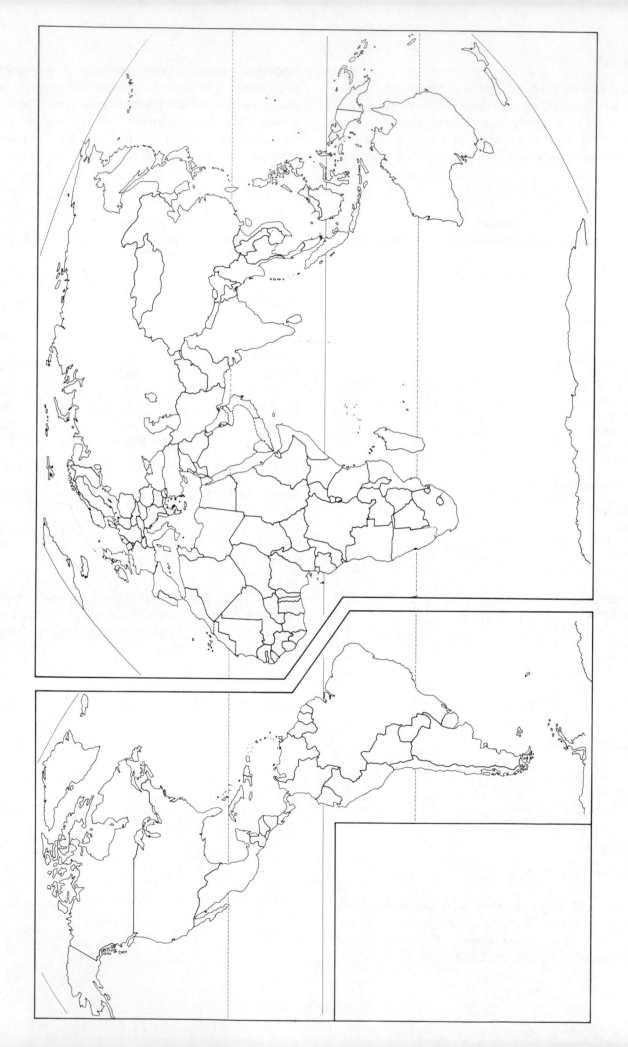

Answer Sheet—1
Wheat—Production, Consumption and Movement

1. Which nations or regions would you expect to be major net exporters of wheat? What demographic and economic characteristics do these nations or regions share?

2. Of the nations listed in Table 7, what percentage of the world's wheat surplus is produced in North America?

3. Which three nations or regions are leading producers of wheat?

4. Which nations or regions listed in Table 7 are the leading consumers of wheat? Why are these nations' or regions' consumption rates so high?

5. Which nation or region is the leading importer of wheat? What percentage of the total imported wheat does this represent?

6. What percentage of the wheat consumed in Africa and the Middle East is actually grown in Africa and the Middle East?

7. Approximately what percentage of the world's wheat harvest would you expect to be exported?

8. Using the population data in Section 5, and the wheat consumption data in Table 7, complete the table (on reverse side of this page) showing the per capita wheat consumption in selected regions.

Consumption of Wheat Per Capita

Region	A 1982-83 Consumption (Million Tons)	B 1983 Population (Millions)	C Lbs. Wheat/Capita/Yr. (A÷B x 2000)
U.S.		234	
Canada		25	
Australia		15	
EC-10*		272	
Argentina		29	
U.S.S.R.		272	
China		1023	
Eastern Europe** (plus Albania & Yugoslavia)		137	
Other Europe***		83	
Brazil		131	
Mexico		76	
Japan		119	
India		730	
Egypt		46	

*Belgium, Denmark, France, Greece, Ireland, Italy, Luxembourg, Netherlands, U.K., W.Germany.

**Bulgaria, Czechoslovakia, East Germany, Hungary, Poland, Romania

***Finland, Iceland, Norway, Sweden, Austria, Switzerland, Malta, Portugal, Spain

Answer Sheet—2
Rice—Production, Consumption and Movement

9. Which 3 nations or regions are the major net exporters of rice?

10. How does the volume of rice surplus among the three leading producers compare with the volume of surplus wheat producers? Which nations or areas appear on both lists of major surplus producers?

11. What percentage of the world rice production is from the United States?

12. Which nation or regions produce a rice surplus? What is the volume of that surplus? What percentage of the world's rice surplus is from the leading two producers?

13. What percentage of the world's rice surplus is from the leading two producers?

14. Which nation is the leading producer of rice? Which is second?

15. Which 3 nations or regions would you expect to be the leading net importers of rice?

16. Approximately what percentage of the world's rice harvest would you expect to be exported?

17. Using the population data in Section 5 and the rice consumption data provided in Table 8, complete the table (on the reverse side of this page) showing the per capita consumption of rice in selected regions.

Consumption of Rice Per Capita

Region	A 1982-83 Consumption Million Tons	B 1983 Population (Millions)	C Lbs. rice/ capita/yr. (A ÷ B x 2000)
United States		234	
China		1023	
India		730	
Japan		119	
Australia		15	
Iran		43	
Nigeria		84	
World		4677	

EXERCISE 3

Answer Sheet—3
Coarse Grains—Production, Consumption and Movement

18. List the 3 countries which you expect to be the major exporters of coarse grains.

19. What percentage of the world's production of coarse grains is from the United States?

20. Which two countries are likely to be the leading importers of coarse grains? Account for why each country imports so much coarse grains.

21. What is the per capita consumption of coarse grains in the United States? What is the per capita consumption in China? What reasons can you suggest to explain this disparity?

EXERCISE 3

Answer Sheet—4
Livestock—Production, Consumption, and Movement

22. Using the world statistical data provided in Section 5, describe the level of affluence of the three leading meat-consuming nations.

23. Which country is the leading per capita consumer of beef and veal?

24. What percentage of the world's population is represented by the major consumers listed in Table 10?

25. Rank the major consumers from most to least meat consumed.

Section 2 Patterns of Economic Development

INTRODUCTION

The long history of human development is marked by many watershed events which signal significant new directions in the progress of humankind. Consider the example of the impact of the Agricultural, Industrial and Scientific Revolutions on demographic change. In a broader context, the four centuries beginning in 1500 were destined, by virtue of the tremendous economic, social, political, technological and intellectual upheavals which characterized them, to drastically alter the fundamental regional and cultural patterns of human occupance of the globe. Consider the events that marked these centuries: 1) the Age of Discovery; 2) the Colonial Era that followed; 3) the birth and growth of nationalism; 4) the expansion of agricultural and food production potential which came as a consequence of the opening of vast new agricultural lands, advances in biological sciences and more efficient farm machinery; and finally, 5) the combination of the Industrial, Scientific, Demographic and Technological Revolutions. These are, individually and collectively, watershed events in human history. That most of these events were perpetrated by, and the benefits accrued primarily to, European cultural areas, is a point vastly significant to full appreciation of current day regional patterns.

Before 1500 A.D., the major cultural regions of the world existed, except for the periodic ebb and flow of peripheral conquests, in relative isolation. By definition, the term culture implies sufficient isolation to foster characteristic behavior patterns unique to one set of people. Today's flash points are often where two or more cultures are competing for dominance: Israel/Palestine; South Africa; Iran; the Indian Subcontinent.

After 1500 A.D., as European cultures spread their influence, the relative balance among cultural regions was drastically altered and the isolation of cultural groups gradually ended. Using the fruits of scientific and technological innovation and, more often than not, with superior, more tightly-woven national organizations, European cultures came to dominate politically, culturally and economically all but a relative handful of the world's cultural regions. In the process, European cultural regions and those places such as Japan, where the innovations of 1500 to the present were successfully diffused, were the primary beneficiaries of the economic and social rewards of the past four centuries. Much of the remaining world is still struggling to assimilate the fruits of the scientific and technological revolutions. The differences between the two groups are marked. The "beneficiaries," representing approximately one-quarter of the world's population, are relatively wealthy, healthy, literate, well-fed and free from want. Among the remaining three-quarters of the world's population, there are significant variations. There are those world regions, nations, and within nations, segments of society, who live at the very margins of existence, often in total degradation. This latter group may number one billion or more of humankind. Finally, there are those who lie somewhere between depravity and wealth. A few nations or societies with the necessary human and physical resources have begun to capitalize on the rewards of scientific and technological advances and evidence some progress towards economic development. Some nations have made use of the wealth generated by abundant resources to make social and economic progress, although not all within the nation necessarily benefit equally. Other nations, burdened by exploding populations, lack of capital and underdeveloped human and physical resources, are still struggling to recognize the benefits of the last four centuries of innovation. Clearly, the benefits of the watershed events of the past four centuries have not diffused successfully or equitably to all areas of the globe. What remains is a patchwork quilt pattern of differential levels of national and regional development.

In search of simplified explanations for the differences among nations, it is convenient to resort to dichotomous measures like: rich-poor, developed-underdeveloped; haves-havenots; or even, us-them. While such measures may prove comfortable, they are rarely sufficient, especially in an academic setting where specific cases—nations, regions, cultures— are to be examined in detail. The natural outcome of detailed scrutiny of any set of places is comparison. In such comparison a more flexible, detailed scale is normally required. A continuum is often more appropriate.

THE DEVELOPMENTAL CONTINUUM

The linkage between economic development and changes in demographic characteristics of a population were outlined in Exercise 1. Less clear is the relationship between industrialization and economic development. Resource wealthy, mostly oil-exporting, nations obtain levels of economic development and standards of living comparable to wealthy, developed, industrialized states without manifesting comparable levels of industrial development. Indeed, some oil-exporting states evidence little industrial development. What then is the nature of this relationship?

One suggested model of the relationship between industrialization and economic development is that posed by economist Walter Rostow. He suggests that there are five stages of economic growth and industrialization is an integral ingredient in the movement of economy through the middle stages:

It is possible to identify all societies in their economic dimensions, as lying within one of five categories: the traditional society, the preconditions for take-off, the take-off, the drive to maturity and the age of high mass consumption. (Rostow, 1971)

In the traditional society, land—the source of most wealth, prestige and income—is in the hands of the landed elite, who occupy the top of a hierarchial social structure. In this predominantly agricultural setting, there is relatively little incentive, nor often the means, to generate large surpluses or to increase productivity. While increases in output are not precluded, they often require innovations to be introduced in trade, industry or agriculture.

For take-off to occur, a high rate of savings and the incentive to reinvest profits and gains in the economy are essential. Rostow emphasizes that it takes time to transform a traditional society in the ways necessary to enjoy the fruits of modern science. The idea that economic progress is possible, even necessary, must spread. Enterprising investors, lured by potentially large profits, must be enticed to invest—particularly in transportation, communications and raw materials. Until the take off stage the pace of economic activity will still be gauged by traditional society values and social structure, and only in the take-off stage when the economy begins to move are old barriers overcome.

In the take-off stage, the economy becomes dominated by growth and expansion, and according to Rostow, "Growth becomes a normal condition." There is a rapid expansion of industry and the transportation infrastructure. During this stage, reinvestment is normal and acts to accelerate economic growth. Growth breeds growth as profits are plowed back into industry, requiring new sources of labor and further expansion of support services. Agriculture is commercialized and the increased productivity necessary for take-off is attained. Innovation becomes acceptable, even necessary, in the search for improved methods of production to expedite profit and improve competitiveness.

As the economy matures, reinvestment is sustained. Modern technology is extended "over the whole front of its economic activity"—i.e., to all industries. New industries emerge as old ones level off. The narrow set of industries that emerged during take-off is expanded usually into higher technology industries. The most advanced innovations science has to offer are, or could be, applied to the entire range of industries developed in the economy.

In the final stage of high mass-consumption, reinvestment, no longer necessary for the mature industrial sector, is reduced. The surplus can now become the means to raise living standards as the economy shifts to durable consumer goods. Some societies cease to accept the further extension of modern technology as the primary objective, choosing instead "to allocate increased resources to social welfare and/or security."

As this brief summary of Rostow's (1971) model indicates, economic development consists of five stages and represents a continuum along which a nation, a region, a society or an economy may progress. In one form or another, the developmental continuum is a convenient and frequently used device, especially in efforts to categorize, for ease of comparison, nations, regions, societies, or economies. The developmental continuum posed by Rostow is but one of many such devices; others make use of different labels for stages of development. Figure 2 summarizes a few of the more commonplace schemes for dealing with variations in levels of development. Can you identify the one suggested by your instructor or in your textbook?

The critical element of any such continuum is utility. The actual differentiation between nations, economies, societies or regions must be based on evidence of economic development. In point of fact, methods for measuring levels of development are relatively poorly developed. Perhaps the methods lack sophistication because the data on which economic development must be based is often suspicious, absent, incomplete and, in too many instances, unreliable. It is often difficult sometimes impossible, to obtain the detailed data on investments, savings and other indicators of economic change required to effectively operationalize Rostow's model. More commonplace are efforts to categorize states by their stage of development, using what amounts to surrogate—and more readily available—indices of development. In this section, we explore several alternative means of gauging levels of economic development.

The three exercises in this section progress from a simple, one-dimensional index to successfully more complex and detailed approaches to measuring development. In Exercise 4, you make use of a single index of development—per capita Gross National Product—to classify 40 nation states by level of development. In Exercise 5, you are asked to employ a set of data which describes various elements of the labor force in order to examine the structure of 40 national economies and to categorize each into one of three levels of development. In Exercise 6, you are provided much more detailed social, economic and demographic data for fewer nations in order to develop a classification system of your own choosing. Finally in Exercise 7, you are introduced to the concept of diffusion, the process by which the new ideas and innovations, which Rostow states are so critical to development, are moved and carried from one place to another.

Figure 2
Comparison of Developmental Continuums

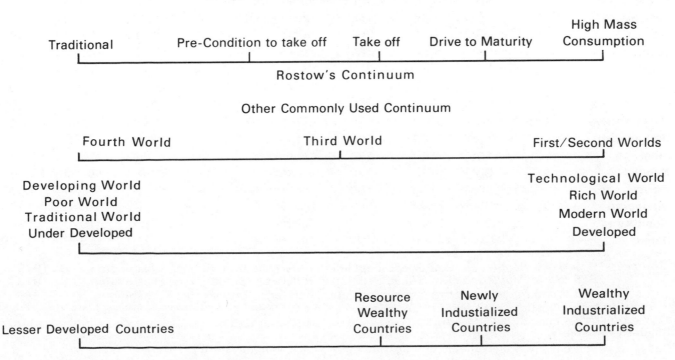

EXERCISE 4

The Wealth of Nations: Using GNP as an Index of Development

In this exercise, you are asked to categorize a set of selected nations according to a single measure of wealth, the Gross National Product (GNP), which is often used as a means of comparing nations at varying stages of economic development. The GNP measures the total domestic and foreign output claimed by residents of a particular country. It is the Gross Domestic Product (GDP), the total output of goods and services produced by an economy, plus income accruing to residents from abroad, less the income earned in the domestic economy accruing to persons abroad. Dividing the GNP by the population of a nation provides a per capita index of economic strength useful in comparing nations with different population sizes.

In this exercise, you are provided the measures of

GNP for 40 countries for a time period covering 10 years from 1973 to 1983. On the graph provided, you are to plot and label a point representing the GNP of each of the countries for each of the three years— 1973, 1978, and 1983. The three points for each country should be connected to establish a trend line, which is a rough measure of economic progress in the 10 years from 1973 to 1983.

You should carefully complete the graph, and then respond to the questions on the next page. Please note that the data for countries with less than 1000 per year per capita GNP is densely clustered. The dots near the bottom line ($0.00 per capita GNP) will be very close together. You should plot them carefully with a sharpened pencil.

Table 11

GNP Per Capita for 40 Selected Countries[1]

Country	Gross National Product/Capita[2]			Country	Gross National Product/Capita[2]		
	1973	1978	1983		1973	1978	1983
Egypt	210	280	700	South Korea	250	670	2010
Spain	1020	2920	4800	Mexico	670	1090	2240
Ghana	310	580	320	Haiti	110	200	320
Mali	70[4]	100	150	Bolivia	180	390	510
Nigeria	120	380	760	Brazil	420	1140	1890
Upper Volta	60	110	180	Venezuela	980	2570	4100
Ethiopia	80[4]	100	140	Argentina	1160	1550	2030
Angola	300	330	790[4]	Denmark	3190	7450	11,490
South Africa	760[3]	1340	2450	Sweden	4040	8670	12,400
Israel	1960	3920	5360	West Germany	2930	7380	11,420
U.S.S.R.	1790	2760	6350	Canada	3700	7510	12,000
Saudi Arabia	440	4480	12,180	East Germany	2490	4220	7286[4]
North Yemen	80	250	510	Romania	930	1450	2546[4]
Bangladesh	70[4]	110	130	Italy	1760	3050	6350
India	110[4]	150	260	Yugoslavia	650	1680	2570
Nepal	80[4]	120	170	Australia	2820	6100	10,780
Burma	80	120	180	United States	4760	7890	14,090
Singapore	920	2700	6620	China	160[4]	410	290
Hong Kong	970	2110	6000	Taiwan	390	1070	2360[4]
Japan	1920	4910	10100	Turkey	310	990	1230

Sources: *1973 World Population Data Sheet,* Population Reference Bureau, Inc., Washington, D.C., 1973.
1978 World Population Data Sheet, Population Reference Bureau, Inc., Washington, D.C., 1978
1985 World Population Data Sheet, Population Reference Bureau, Inc., Washington, D.C., 1985.

[2] In U. S. Dollars.
[3] Includes Namibia.
[4] Cases in which data was unavailable and previous year's data was substituted instead.

EXERCISE 4

Answer Sheet

1. Which eight countries had the greatest GNP in 1983? Were these same eight the "wealthiest" in 1973? Which were different?

    Saudi Arabia   West Germany
    Japan   Canada
    Denmark
    Sweden

2. Of these eight nations, which one seems to be an anomaly? How is it different? What might explain its recent economic history?

3. Which eight nations had the lowest per capita GNP in 1983? What are some of the characteristics these nations have in common? Have the per capita GNP figures of these nations increased much from 1973 to 1983? Why or why not?

4. Compare the eight wealthiest nations (from question 1), and the eight poorest nations (from question 3). What differences do you see in their rates of growth. Will the poorer nations ever catch up to the GNP incomes of the wealthier nations? Why or why not?

5. Using 1983 data, select the eight countries with per capita GNP figures between $3000 and $8000. What characteristics do these eight nations have in common? What were their rates of growth like over the period between 1973 and 1983?

6. If you use 1983 data and observe the clusters of dots along the vertical line representing 1983, you can spot several groups of nations. Natural breaks or gaps appear between clusters of dots. Three such clusters might be: above $10,000; $4000-$7500; and less than $3000. Are three groups adequate? Can you further subdivide each group? What type of data would you need to complete a more finely divided classification system?

7. If you had to make a projection about the future economic growth of each of your groups, how would you expect them to change during the next five years. Based on the three years of data you already have, project what the per capita GNP will be in 1988.

8. Is per capita GNP an adequate measure of development? Offer support for your answer. What are the strengths and weaknesses of per capita GNP as a measure of development?

EXERCISE 5

The Structure of a National Economy: Assessing Economic Development with Multiple Labor Force Indices

Assessing the level of economic development of a nation is critical to a full appreciation of the nature of that place. The degree of deprivation or affluence of the inhabitants of a place is a function of the level of economic development. Unfortunately, index measures of development like per capita GNP present some problems when used as a means of analysis. A single measure, particularly a measure of total wealth, can mask the true character of a national economy. When national income and wealth are concentrated in the hands of a relatively small minority, who control the production and income-generating functions of the nation and the vast majority of the population receives little income, then a general measure of wealth, like per capita GNP, can prove insufficient to gauge development. If the overwhelming majority of the population are rural, traditional agriculturalists, who are relatively unaffected by the national economy, then in Rostow's terms, that society is still a traditional society. Unfortunately, the per capita GNP may rival that of the wealthier industrialized nations.

To overcome the shortcomings of such general measures of national or personal income, it is often necessary to determine how the economy is structured. Is there an industrial sector of any consequence? How strong are each of the other components of the economy such as the agriculture and service sectors?

The structure of a national economy is frequently analyzed by utilizing data on how the population makes their living. The employment structure is analyzed by dividing the labor force into three mutually exclusive groups: primary activities such as mining, forestry, and agriculture; secondary activities including industrial and manufacturing activities, and service activities, which in the broadest sense includes wholesaling, retailing, transportation, business and consumer services, education and government. The proportion of the work force employed in each of these activity sectors offers a reasonable indication of the type of economic development in a given country. Generally, the smaller the proportion of workers in the primary sector, the more advanced the economy. Conversely, nations or regions with a high percentage of the work force in the services are sometimes referred to as "post-industrial"—Rostow's high mass consumption—societies. Just as Europe and the United States shifted from a rural agricultural society to an industrial society during the 19th century, some nations, during the 20th century, seem to be shifting away from manufacturing to high technology jobs in computer design, mathematics, economics, urban and regional planning, and other scientifically based knowledge industries.

Using agricultural employment as a surrogate for employment in the primary sector, Table 12 provides basic data on the mix of employments in each of the three sectors (primary, industrial, and service) for the same 40 national economies you have previously examined. On the triangular graph on page 43, locate the point, for each of the 40 countries, which corresponds to the proportional mix of the labor force employed in each of the three sectors.

Each point on your graph will represent one country. Each point will be the product of three separate statistics: percentage of the labor force employed in industry; in agriculture; and in the service sector. The value for each statistic is plotted from each separate side of the triangle. Where the three plots intersect is the overall value for the national economy.

Locating all the points within the triangle makes it easier to compare and contrast the relative stages of economic development among all 40 nations. It is possible to develop categories of economic development. Using your final plot of nations, group the 40 nations into three distinct classes of economies: agricultural, industrial, and service. (Suggestion: Divide the graph into three equal sections by drawing lines perpendicular to the triangle's three sides. From point a on side BC, point b on side AC and point c on side AB, draw three lines that will intersect in the center of your graph.) Once you have plotted all the data, use your graph to respond to the questions on the answer sheet for Exercise 5.

Table 12

Labor Force Characteristics for Selected Countries

Country	Agriculture	Industry	Services	Country	Agriculture	Industry	Services
Egypt	50	30	20	Japan	12	39	49
Spain	15	40	45	Korea, South	34	29	37
Ghana	53	20	27	Taiwan	34	37	29
Mali	73	12	15	Canada	5	29	66
Nigeria	54	19	27	United States	2	32	66
Upper Volta	82	13	5	Mexico	36	26	38
Ethiopia	80	7	13	Haiti	74	7	19
Angola	59	16	25	Bolivia	50	24	26
South Africa	30	29	41	Brazil	30	24	46
Israel	7	36	57	Venezuela	18	27	55
Saudi Arabia	6l	14	25	Argentina	13	28	59
Turkey	54	13	33	Denmark	7	35	58
Yemen, North	75	11	14	Sweden	5	34	61
Bangladesh	74	11	15	Germany, West	4	46	50
India	69	13	18	Germany, East	10	50	40
Nepal	93	2	5	Romania	29	36	35
Burma	67	10	23	Italy	11	45	44
Singapore	2	39	59	Yugoslavia	29	35	36
China	71	17	12	Australia	6	33	61
Hong Kong	3	57	40	U.S.S.R.	14	45	41

Source: *World Development Report, 1983,* Oxford University Press, London, 1983.

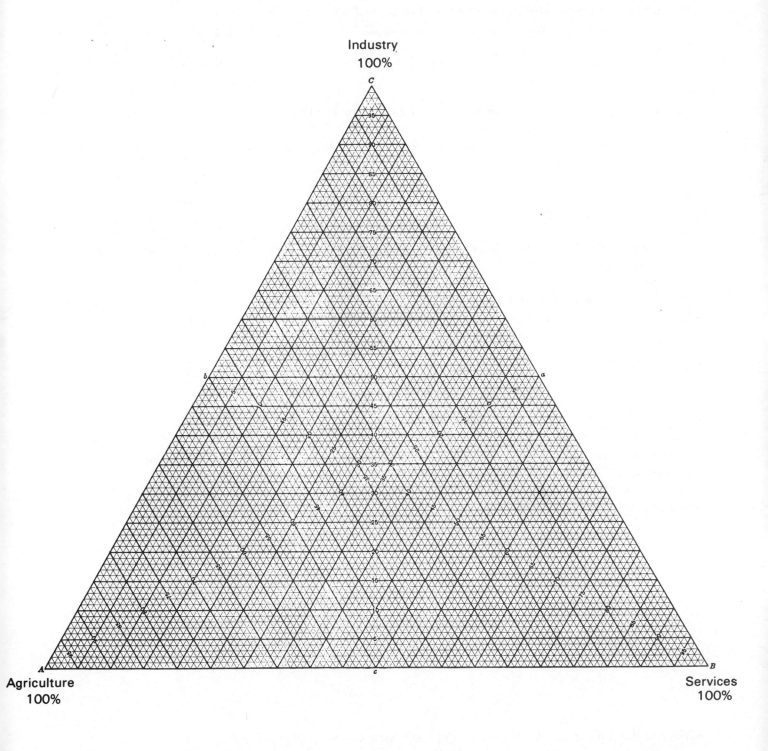

Industry
100%

Agriculture
100%

Services
100%

EXERCISE 5

Answer Sheet

1. Which countries have economies dominated by agriculture?

2. Which five countries have the smallest percentage of their labor force employed in agriculture?

3. Of those countries dominated by agriculture, which has the highest percentage of its labor force in industrial jobs?

4. Which two Latin American nations appear most likely to be transitional economies on their way to becoming modernized nations?

5. Daniel Bell, in *The Coming of the Post-Industrial Society*, proclaims the United States to be the first such society. From the triangular graph, list three other countries that are close to the United States and might also be considered post-industrial.

6. Which four countries have the greatest percentage of the labor force employed in industry?

7. Compare this method of classifying nations in developmental stages to the per capita GNP approach used in Exercise 4 and to any one of those mentioned in Figure 2 of the Introduction to Section 2. In making this comparison, it might be best to use the developmental continuum employed by your text. Is the use of labor force characteristics a valid method of classifying the level of development of nations? Offer support for your response.

8. What are the advantages of this method of classifying nations? Disadvantages?

EXERCISE 6

The Structure of a National Economy: Assessing Development Using Detailed Data

In the two previous exercises you were asked to compare the national economies of 40 countries by using overall measures of development. Both GNP per capita (Exercise 4) and the labor force measures (Exercise 5) were intended to serve as surrogates of economic development. As is often the case, surrogates have shortcomings, not the least of which, in this instance, is that they are one dimensional. Both the GNP and labor force indices characterize single dimensions of economic development. While each, in its own way, may suffice for quick and relatively simple estimates of economic development and/or for comparing the economies of large numbers of countries, any accurate estimate of the stage of development of a given nation will depend on a more detailed description of multiple facets of national well-being. Economic measures alone may prove misleading. In order to determine if the wealth of an economy is being transferred to all segments of society, it is important to gauge social, as well as economic, progress. For example, it is important to estimate the accessibility that the general population has to key elements of individual well-being, such as medical services (for adults and infants), social services, education and sufficient food to sustain a healthy existence. It is quite possible for a nation to rank high on a per capita GNP measure and low on indicators of social well-being. It is also quite possible for a country to rank fairly low in per capita GNP and high on social welfare indicators.

Measures of social well-being are quite illustrative of development. The question is, once again, how to accurately and efficiently assess social well-being. There are several options.

Summary measures of social well-being exist. The Quality of Life Index (QLI) measures standard of living based on key medical, literacy and social variables. However, QLI is open to the same criticisms as any single, one-dimensional index.

In most circumstances any assessment of development is best done by fairly detailed economic and social indicators. However the use of a large amount of information on individual countries precludes the comparison of large numbers of nations. There are statistical procedures which allow a researcher to mathematically reduce dozens of variables portraying the detailed social and economic characteristics of hundreds of countries to a few overriding dimensions or clusters of variables. Using these clusters the researcher may then group or categorize nations according to the rank of each on the various clusters or dimensions. However, such statistical procedures require considerable training and experience to be effective; hence, their use and usefulness is thus confined to a fairly specialized or professional setting or clientele. Thus for all intents and purposes large data sets for large numbers of countries are not generally feasible for an introductory level course such as the one in which you are enrolled. But for small groups of nations it is quite feasible to scrutinize fairly detailed sets of information in order to estimate levels of social and economic development. In this execise you are asked to accomplish just such a task.

In this exercise you will make use of a data set consisting of 30 variables for 10 countries (Table 13) in order to construct your own developmental continuum. Use the data provided in Table 13 to determine a ranking from the least-developed to the most-developed of the 10 nations for which data is provided. Draft a scale such as those found in Figure 2 (page 35) of this section. Determine what position each of the 10 nations should occupy on your scale. Use your scale and the data in Table 13 to respond to the questions on the next page.

Table 13

Indices of Economic Development**

	ANGOLA	BRAZIL	ETHIOPIA	EAST GERMANY	MALI	ROMANIA	SAUDI ARABIA	SINGAPORE	SWEDEN	YUGOSLAVIA
*GNP Per Capita	790	2,214	142	7,286	185	2,546	12.720	5,220	14,500	2,789
*Percent Urban Population	21	68	14	76	17	49	70	100	83	39
*Per Capita Calorie Supply as Percent of Requirements	89%	105%	74%	142%	84%	128%	119%	—	117%	138%
*Population in Millions	7.6	131.3	31.3	16.7	7.3	22.7	10.4	2.5	8.3	22.8
*Birth Rate/1000	47	31	48	14	47	18	44	17	11	17
*Death Rate/1000	22	8	23	14	21	10	13	5	11	9
*Infant Mortality/1000	153	76	146	12.3	153	29.3	112	11.7	7.0	30.6
*Doubling Rate (in Years)	27	30	27	—	26	91	22	58	3465	90
*Natural Rate of Increase	2.5	2.3	2.5	0.0	2.6	.8	3.1	1.2	0.0	.8
●Arable Land Per Capita (in Acres)	.5	.7	1.1	.7	3.8	1.1	.3	.002	.9	.8
#Percent Labor Force in Agriculture	59	30	80	10	73	29	61	2	5	29
#Percent Labor Force in Industry	16	24	7	50	12	36	14	39	34	35
#Percent Labor Force in Services	25	46	13	40	15	35	25	59	61	36
×Area (mi² × 1000)	481	3,286	472	42	479	92	873	.238	173	99
×Population Density (in persons/mi²)	15.5	37.9	70.8	41.2	15.8	245.7	11.1	10,504.2	52	226.9
●Physicians Per 100,000	6	68.1	1	190	5	135	118	78	178	131
●Teachers Per 1000	9	25	3	43	4	30	22	21	46	26
●Literacy Rate (Percent)	20	68	8	99	10	98	15	84	99	85
●Hospital Beds Per 100,000	306	327	29	1065	56	919	155	371	1496	603
×Imports in Billions (U.S. $)	.625	936	.787	20.196	.332	9.836	40.654	28.167	27.591	13.346
×Exports in Billions (U.S. $)	1.227	18.627	.404	21.743	.146	11.714	79.123	20.788	26.817	10.265
●Beef Production in 1000s Metric Tons	51	2500	214	370	35	320	19	—	160	343
●Newspaper Circulation Per 1000	16	44	1	515	—	181	29	264	578	97
●Radios (in Millions)	125	35	.25	6.4	.082	3.2	2.7	.608	8.3	4.8
+Energy Consumption	255	1101	24	7412	31	4775	6764	8544	7971	2402
+Average National Savings Rate	18.9	19.9	5.7	—	3.9	—	61.8	27.2	16.7	29.9
+Food and Beef Imports	—	6.8	6.2	—	19.1	—	13.7	11.5	8.8	9.0
+Fuel Imports	—	34	15.1	—	14.3	—	.6	25.6	17.4	13.4
#Percent of GNP in Domestic Investment (1980)	9	20	10	—	16	33	26	42	19	32
#Percent of GNP in Domestic Savings (1980)	19	19	4	—	-6	—	59	33	18	29

Sources:

● *The World Almanac and Book of Facts, 1985,* Newspaper Enterprise Association, Inc., New York, 1985.

* *1985 World Popualtion Data Sheet,* Population Reference Bureau, Inc., Washington, D.C., 1985.

× *Information Please Almanac Atlas and Yearbook, 1985,* 38th edition, Simon and Schuster, New York, 1985.

+ *World Tables,* 2nd edition, 1980, from the data files of the World Bank, Johns Hopkins University Press, Baltimore, 1980.

World Development Report, 1983, The World Bank, Oxford University Press, London, 1983.

** Some data is taken from earlier editions of the sources listed; in each instance the data are the most recent.

EXERCISE 6

Answer Sheet

1. After careful study of the data in Table 13, use the line provided below to draft your own developmental continuum. Position each of the ten nations somewhere along your continuum.

2. List below the variables which you feel were most important in arriving at your final ranking of these ten nations. In other words, how and why did you rank each nation where you did?

3. Now rank these nations according to:

    a) GNP per capita (lowest to highest)

    b) percent of the total labor force in agriculture (lowest to highest)

Compare these rankings with the rankings in your continuum. Explain the differences in the three rankings.

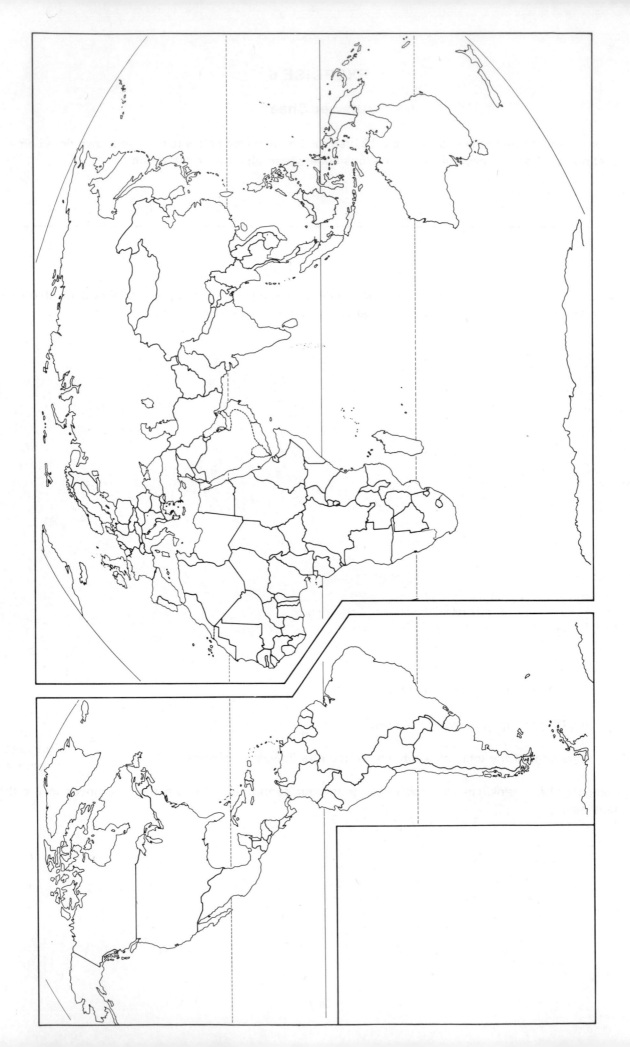

EXERCISE 7

The Spatial Diffusion of Innovation

The character of nations and regions differs widely between the most developed industrial areas and the isolated, pre-agricultural cultures. Whether economic, political, linguistic, religious, or social, most of these differences result from variations in accessibility and receptability to new ideas and innovations. The movement of ideas and innovations across areas is called spatial diffusion. This process is of great interest to geographers. The role of diffusion and acceptance of new scientific and technological ideas is *sine qua non* in the transition of a national economy from a traditional society through take-off to a mature industrial one. As Rostow himself states:

> New techniques spread in agriculture as well as industry, as agriculture is commercialized, and increasing numbers of farmers are prepared to accept the new methods and the deep changes they bring to ways of life.

Every culture trait from agriculture, pottery, and writing, to microprocessors, natural gas pipeline turbines, AIDS, and the popularity of Duran Duran have patterns of spatial diffusion. The diffusion pattern for any of these traits results from a complex interaction between "barriers" which resist the spread of the trait and "carriers" which facilitate the spread of the trait. The purpose of this exercise is to introduce you to the mechanics of spatial diffusion. The problem to be solved is a hypothetical one, but it is representative of many examples of diffusion.

The map represents a volcanic island in the western Pacific Ocean inhabited by peoples living in rural villages. Each village lies near the center of a five square mile area containing small agricultural fields or tropical rainforest. The island is politically divided and the villages in the northwest corner of the island are isolated from communicating with the other islanders by armed border guards, barbed wire, and mine fields. Most of the interior villages are free farming villages dependent on intensive subsistence agriculture, primarily wet rice cultivation. The volcanic mountains in the southeastern quarter of the island are considered sacred by the indigenous peoples and are therefore uninhabited. The coastal villages are fishing and gathering culture groups that do not practice agriculture. The central island area is inhabited by a conservative religious sect who farm their land but choose not to accept any modern technological innovations.

During the last month, farmers in a southwestern village (find X on the map on the Answer Sheet for Exercise 7) received a free and abundant supply of the herbicide paraquat from the British manufacturer, Imperial Chemical Industries, as part of a special introductory offer. Paraquat is probably the most effective herbicide in existence today. Spraying paraquat is so effective in killing weeds that it can virtually replace much of the tilling and cultivation necessary to kill competing plants. In traditional systems of agriculture, much back-breaking human labor is saved. The perceived advantages to be gained by adopting paraquat are so great that every time a farming village learns about it, they adopt the innovation.

Contact between villages is by word of mouth only. There are no telephones, radios, or televisions. Travel between villages is by ox-cart over dirt trails that weave between the fields. An innovation such as paraquat spreads only by personal contact between the leading male from each village. These leaders normally travel to see the leading male of a nearby village only once each month. Because of the time and effort spent in travel, closer villages are visited much more frequently than more distant villages. Contacts cannot be made behind the political boundary to the northwest. The religious sect will not use paraquat regardless of how often they hear about it and they do not tell anyone about it. Because the fishing villages do not farm, the innovation makes no impression on them and they neither accept the innovation nor tell anyone about it. No contacts can be made at sea. Other than these constraints, contact between village leaders and the diffusion of the innovation is a random process.

Assuming abundant supplies of paraquat, shade in the area of paraquat users ten years ago hence. Assume that the physical and cultural "barriers" and "carriers" remain unchanged during the decade.

Answer Sheet - Map

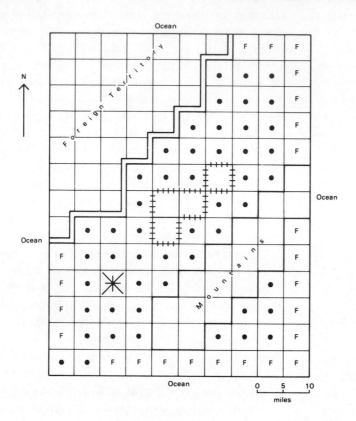

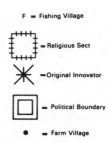

F = Fishing Village

= Religious Sect

= Original Innovator

= Political Boundary

= Farm Village

1. Describe what you would expect the diffusion pattern of paraquat to be after five months.

2. Which factors acted as barriers to the spread of the innovation?

3. In this example, which seems to have been more important, enviromental barriers or cultural barriers? List the enviromental barriers. List the cultural barriers.

4. List other enviromental factors which you think could act as deterrents to the spread of an innovation.

5. List other cultural factors which you think could act as deterrents to the spread of an innovation.

6. What five other cultural traits (or innovations?) do you think would have interesting spatial diffusion histories?

Section 3
Regional Study
Guides

INTRODUCTION

Like many other disciplines geography delights in tracing its academic heritage back to antiquity. And for the two millennia that geographers claim their profession has been practiced, one principal focus has been the study of regions. The regional approach to geography focuses on the identification of a section of the earth and the study of all aspects of that place geography. For an unfortunate number of people, their only exposure to the study of geography may have consisted of memorizing state capitals and other similar minutia in early secondary school geography class. As you have probably learned by now, from either, or both, your instructor or your text, geography is a discipline which has other dimensions than memorizing place names. As you will discover, geography has exciting dimensions beyond regional studies. We encourage you to explore other facets of the profession. However, the mere fact that you have this book suggests that you are enrolled in an introductory world regional course. As such, you are engaged in a survey of selected cultural, economic, political or physical regions of the world.

In any survey course, particularly one that will focus for brief periods on large volumes of material (regions of the world, for example), it is difficult to fully assimilate everything being disseminated. The task is made even more difficult if you are one of the many who have failed to update or refresh your information base on world events, places and locations. If *your* last contact with geography was a secondary or high school class, in the sixth, eighth or even twelfth grade, you may find that, like any tool, geographic information tarnishes with age and disuse. To maximize your opportunity to learn from your classroom experience you may want to refresh your memory about some of the places you will study.

Of course, regardless of whether your last exposure to such geographic information was 6 months, 6 years or 16 years ago, refreshing your memory is a useful effort. With this in mind we have included in this section a series of 9 regional study guides.

The regional study guides are intended to prepare you for the classroom or lecture experience. The terms, places and concepts with which you are asked to become familar are found in most introductory texts, although any given term may not be introduced within the particular regional chapter with which we have associated it. If you prepare for your lecture experience by reading the appropriate chapter in your text *and* by working through the suggested tasks in the study guides, you reduce significantly the chance that you will fail to grasp an aspect (maybe a critical aspect) of the lecture because you were geographically unprepared—i.e., you had no real idea where the places about which the instructor was talking were located. Please understand these terms and places are not simply more state capitals, highest mountains, or longest rivers to memorize. Rather, they are terms and places that are important components in any regional overview and assessment. We think you will find that completing these study guides will prove to be a very useful aid in obtaining an appreciation for, and perhaps even an understanding of, the region in question.

STUDY GUIDE 1
United States and Canada

1. Define and note the significance of:

accessibility	economies of scale
urban sprawl	conspicuous consumption
megalopolis	Standard Metropolitan Statistical Area
hinterland	break-in-bulk points
environmental amenities	Canadian Shield

2. Know the location of the following:

Social/Economic/Cultural Features	Physical/Environmental Features
all states and provinces	Rocky Mountains
Los Angeles	Appalachian Mountains
New York City	Sierra Nevadas
Chicago	Great Plains
Philadelphia	Central Valley of California
Houston	Mississippi River
Boston	Ohio River
Detroit	Missouri River
Toronto	Arkansas River
Montreal	Colorado River
Dallas	Tennessee River
San Diego	Columbia River
Baltimore	Yukon River
San Antonio	Rio Grande
Phoenix	Colorado Plateau
Indianapolis	Columbia Plateau
San Francisco	Long Island
Winnipeg	the Great Lakes
Vancouver	Cascade Mountains

3. Using the map on the reverse side and your textbook as a guide, outline and shade in the major districts within the United States and Canada.

4. Using the map on the reverse side and your textbook as a guide, outline and shade in the major agricultural regions within the U.S. and Canada.

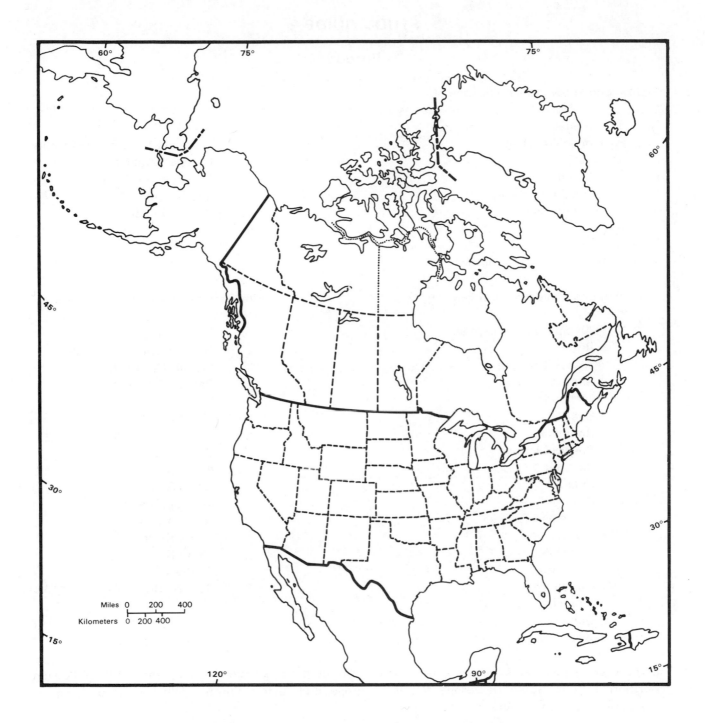

Miles 0   200   400
Kilometers 0 200 400

STUDY GUIDE 2

Europe

I.   Define and know the significance of:

feudalism   NATO
neocolonialism   political and cultural diversity
EEC   European unification
Common Market   European Parliament
complementarity   shatter belt
nationalism   Warsaw Pact

2.   Know the location of the following:

Social/Economic/Cultural Features   Physical/Environmental Features

All countries & capitals   North European Plain
Ruhr   Scandinavian Peninsula
Alsace-Lorraine   Iberian Peninsula
Sambre-Meuse   Balkan Peninsula
Midlands   The Alps
Saar   The Jura Mts.
Bohemia   The Pyrenees Mts.
Silesia-Moravia   Carpathian Mts.
Saxony   Baltic Sea
Milan   North Sea
Lyon   Adriatic Sea
Barcelona   Aegean Sea
Zuider Zee   Rhine River
Marseilles   Seine River
Bordeaux   Po River
Gibraltor   Danube River
Munich
Geneva
West Berlin

3.   Using your textbook as a reference delimit and color on the accompanying map each of the following:
     Northern Europe, Southern Europe, Western Europe, and Eastern Europe. What is the unifying
     characteristic of each region?

54

STUDY GUIDE 3
The Soviet Union

1. Define and note the significance of:

    cultural pluralism   collective farms (kolkhoz)
    collectivization   state farms (sovkhoz)
    permafrost   core triangle
    taiga   centralization
    primate city   the totally planned economy
    tundra   politburo
    virgin lands project

2. List the 5 major ethnic groups in the Soviet Union in order of relative size.

3. Know the location of the following:

    Cultural/Economic/Social Features   Physical/Environmental Features

    All Soviet Socialist Republics   Ural Mountains
    Leningrad   Siberia
    Moscow   Caucasus Mountains
    Kiev   Volga River
    Murmansk   Dnieper River
    Yakutsk   Ob River
    Vladivostok   Caspian Sea
    Gorky   Black Sea
    Irkutsk   Lake Baikal
    Kharkov
    Novosibirsk

4. Using the map on the reverse side and your textbook as a reference, outline and shade in the 5 principal industrial areas of the Soviet Union.

5. Locate the major agricultural regions of the Soviet Union.

6. Using your text as a resource, describe the physical and cultural landscapes you would expect to see during an overland trip from Tashkent through Karaganda, Magnitogovsk, Gorky to Murmansk.

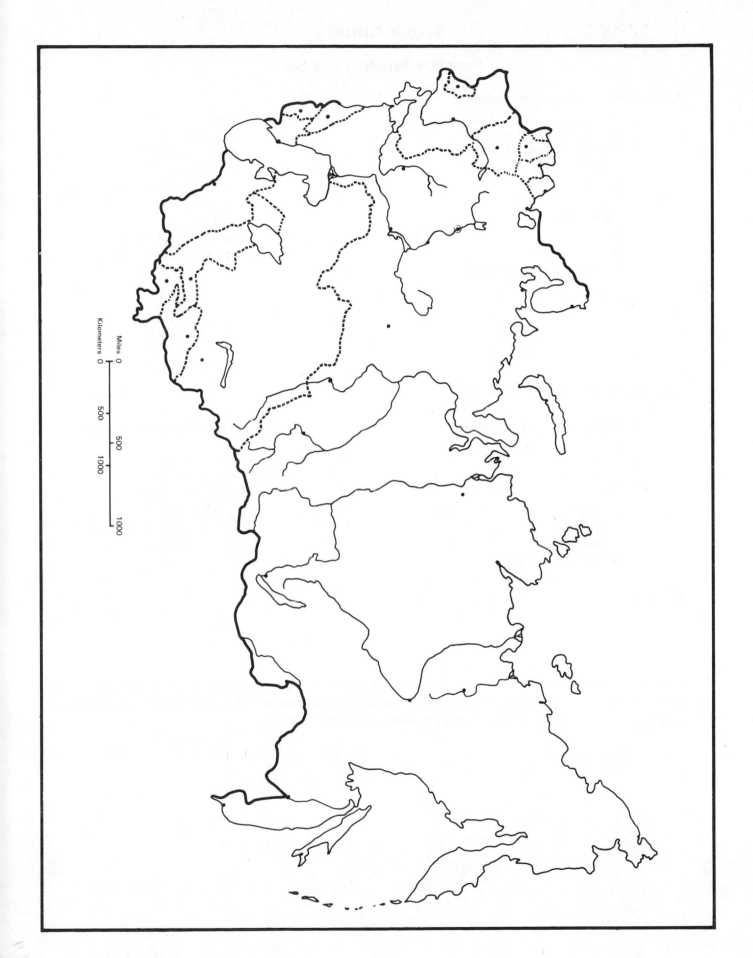

Miles 0

Kilometers 0

500

500

1000

1000

STUDY GUIDE 4

People's Republic of China

1. Define and know the significance of:

The Cultural Revolution
The Great Leap Forward
Mao Tse-tung
Chiang Kaishek
loess
Hong Kong
monsoon
Manchu dynasty
Taiwan (Nationalist China)

alluvial soils
Han Chinese
terracing
the five-guarantees
winter wheat
kaoliang
people's communes
sphere's of influence

2. Know the location of the following:

Social/Economic/Cultural Features

Peking
Manchuria
Canton
Tientsin
Shanghai
Taiwan
Sinkiang Province

Physical/Environmental Features

Hwang Ho River
Yangtze River
Great Plain of China
Gobi Desert
Taklamakan Desert
Tibetan Plateau
Tarim Basin

3. More than three-fifths of the Chinese population are farmers. Using your text as a reference, sketch and label the principal agricultural regions of China on the outline map on the next page. What factors might help to explain the types of agriculture practices in Western China?

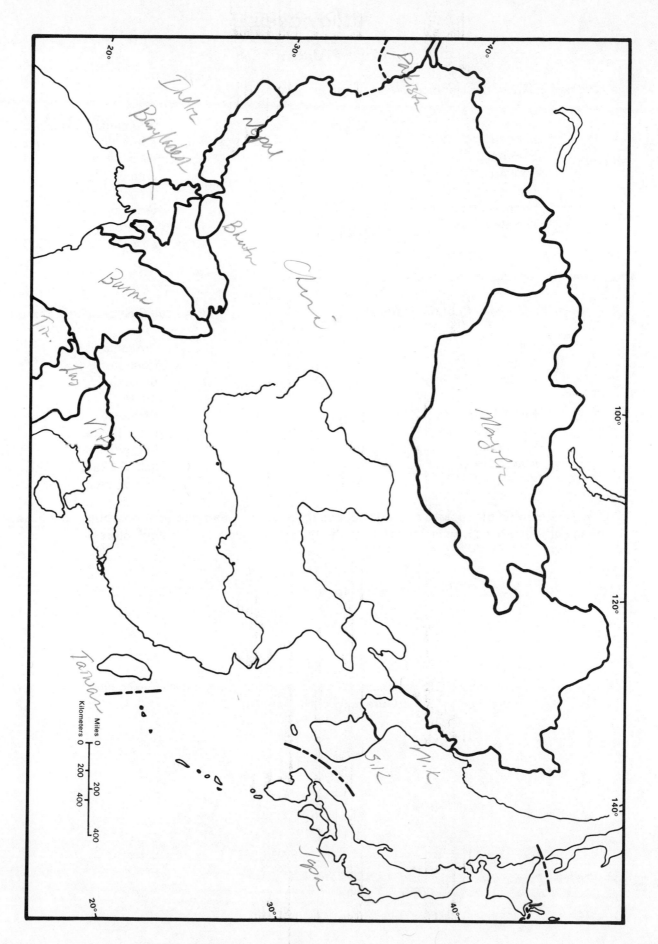

Delhi
Bangladesh
Nepal
Pakistan
Bhutan
China
Burma
Mongolia
Tibet
Lao
Vietnam
Taiwan

Miles 0
Kilometers 0
200
200
400
400

N.K.
S.K.
Japan

20°
30°
40°
100°
120°
140°
20°
30°
40°

59

STUDY GUIDE 5

Japan

1. Define and know the significance of:

typhoons	fragmented farms
shogun	ethnically homogeneous
samurai	social change
insularity	archipelago
tariffs	Ring of Fire
intensive land use	terraces
Zen Buddhism	conurbation
	air pollution

2. Know the location of the following:

Social/Economic/Cultural Features	Physical/Environmental Features
Tokyo	Hokkaido
Nagoya	Honshu
Osaka	Shikoku
Kobe	Kyushu
Hiroshima	Okinawa
Yokohama	Mt. Fujiyama
Kawasaki	Kanto Plain
Kinki District	Inland Sea
Kitakyushu	Sea of Japan

3. Using the map on the reverse side and your textbook or any other reference as a guide, outline and label the major industrial districts according to the type of products each manufactures.

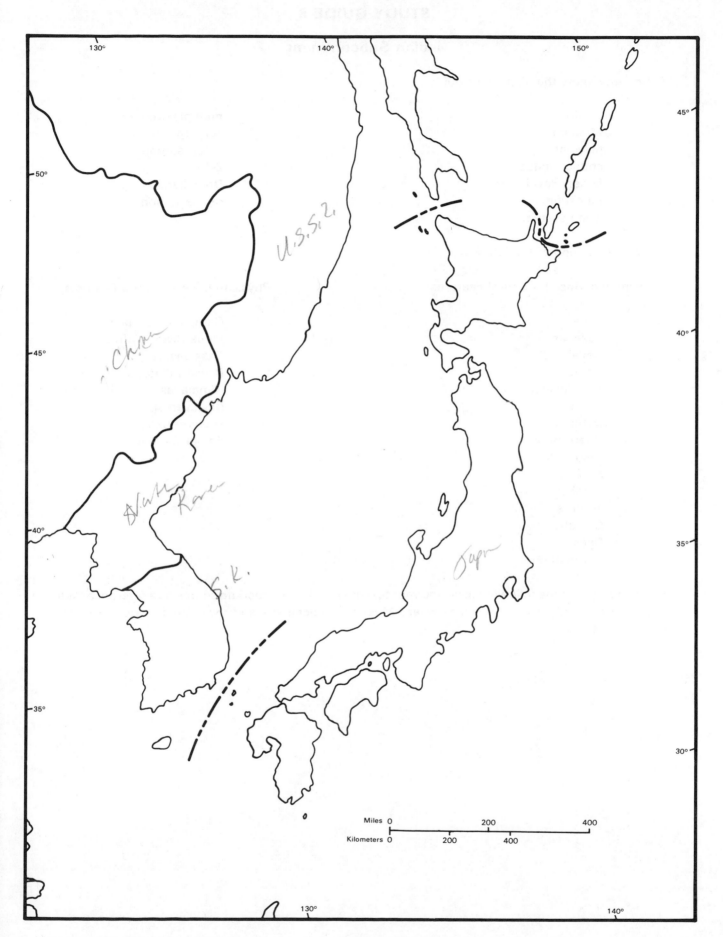

STUDY GUIDE 6

Indian Subcontinent

1. Define and know the significance of:

Hindus   multiple cropping
Muslims   salinization
polyglot   deforestation
cottage industry   Zebu
Green Revolution   Dravidians
monsoon   reincarnations
caste system

2. Know the location of the following:

Economic/Social/Cultural Features   Physical/Environmental Features

India   Ganges River/plain
Pakistan   Indus River/plain
Nepal   Thar desert
Bhutan   Deccan Plateau
Bangladesh   Himalayas
Sri Lanka   Mt. Everest
Delhi   Arabian Sea
Cherrapunji   Bay of Bengal
New Delhi
Karachi
Bombay
Madras
Calcutta
Dacca
Kathmandu

3. Using the map on the following page and your text or any other appropriate source as a reference sketch in and label the outline of those areas where Muslims predominate and those areas where Hindus predominate.

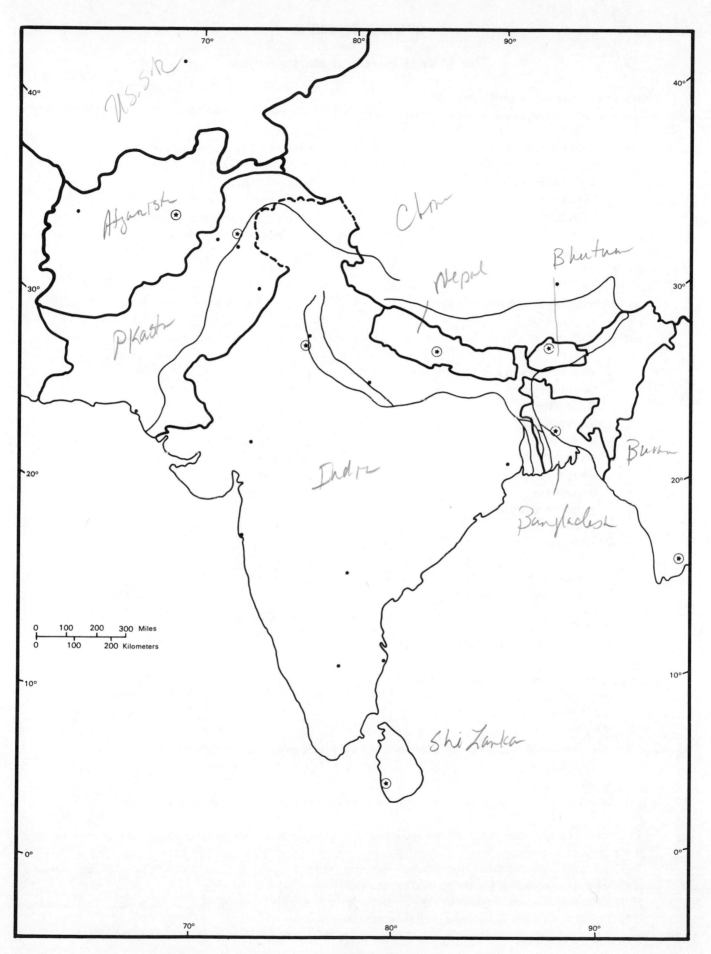

U.S.S.R.

Afganistn

China

Nepal

Bhutan

P.Kastn

Bhutan

India

Bangladesh

0 100 200 300 Miles
0 100 200 Kilometers

Shi Lanka

STUDY GUIDE 7

The Middle East and North Africa

1. Define and know the significance of:

OPEC	Zionism
Mohammed	Ottoman Empire
sunnite muslim	West Bank
shiite muslim	monotheism
Persia	nomadism
Palestine	Balfour Declaration
PLO	McMahon Agreements
Islam	Judaism
	Christianity

2. Know the location of the following:

Social/Economic/Cultural Features	Physical/Environmental Features
all countries in the region	Strait of Hormuz
Golan Heights	Red Sea
Gaza Strip	Bosporus Straits
Amman	Dardanelles Straits
Medina	Sahara Desert
Mecca	Atlas Mountains
Beirut	Arabia Desert
Jerusalem	Tigris River
Western Sahara	Euphrates River
Damascus	Nile River
Teheran	Sinai Peninsula
Kharg Island	Persian Gulf
Aswan Dam	Maghreb
Suez Canal	
Cairo	

3. Where are the majority of the Palestinians located?

4. List the 6 major language groups of the Middle East.

5. List the countries of OPEC. Are all the major oil *exporters* members of OPEC? Which are not?

6. It is probable that one of the following will be a major topic of discussion in your class:

   1) Emergence of Israel, the continuing Israeli/Palestini conflict, the consequences of that conflict, and the prospect for peace;
   2) The conflict precipitated in resource wealthy nations when efforts to offset rapid modernization collides with traditional lifestyles and values (especially Islamic values);
   3) The emergence and impact of OPEC on the economics and politics of this region and the impact of OPEC on the wealthy developed, urban-industrial societies.

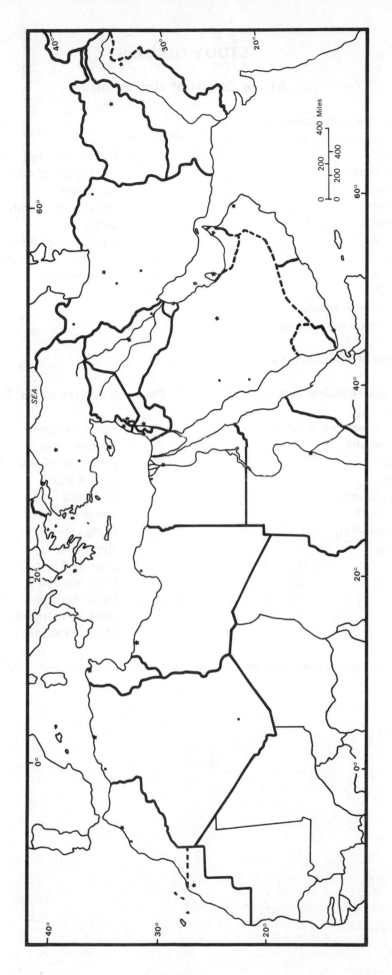

STUDY GUIDE 8

Africa South of the Sahara

1.   Define and know the significance of:

Sahel

desertification

apartheid

tribalism

Pangea

rift valley

steppe

savanna

tropical

pastoral nomadism

subsistence agriculture

cash cropping (commercial agriculture)

the legacy of colonialism

Afrikaans

urbanization without industrialization

tsetze fly

sleeping sickness

shifting agriculture

lateritic soils

2.   Know the location of the following:

Economic/Social/Cultural Features

all the countries of Africa

Addis Ababa

Lagos

Nairobi

Ouagadougou

Cape Town

Johannesburg

Soweto

Physical/Environmental Features

Kalahari desert

Namib desert

Ethiopian Highlands

Congo Basin

The Veld

Zambezi River

Congo River

Niger River

Orange River

Lake Malawi

Lake Victoria

Lake Tanganyika

Mt. Kilimanjaro

3.   On the map on the next page trace the outline of the major climatic regions of Africa.

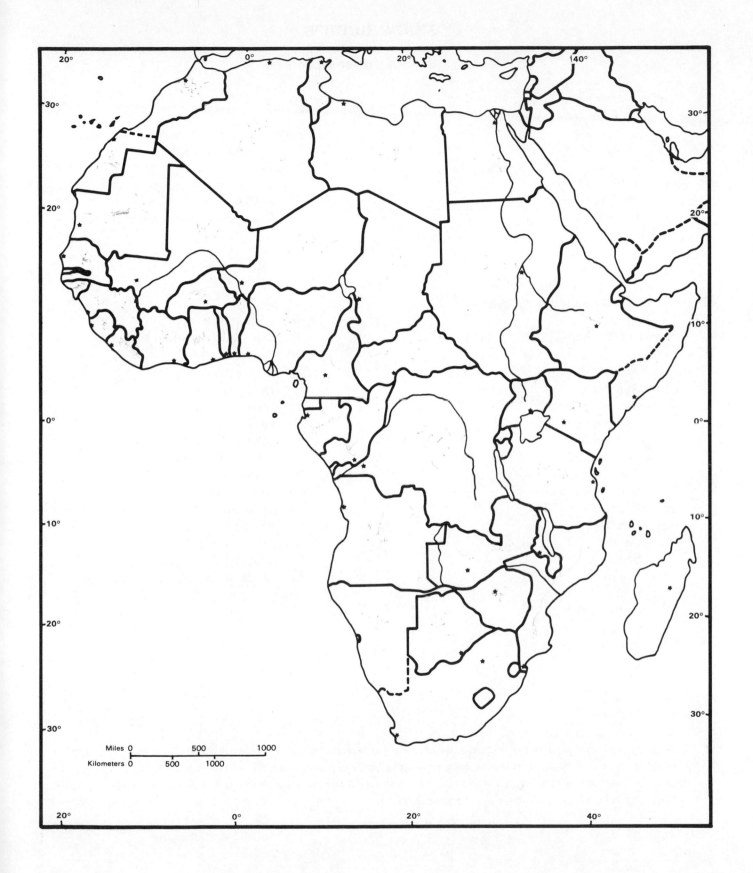

STUDY GUIDE 9

South America

1. Define and note the significance of:

Antiplano

pampas

vertical zonation

rainforest

hacienda

land tenure

barrios

campesinos

LAFTA

Organization for American States

2. Know the location of the following;

Social/Economic/Cultural Features

All the countries of Latin America

Sao Paulo

Rio de Janeiro

Santiago

Buenos Aires

Montevideo

Porto Alegre

Salvador, Brazil

Recife

Fortaleza

Caracas

Bogota

Lima

Quito

La Paz

Falkland Islands

Galapagos Islands

Physical/Environmental Features

Andes

Lake Maricaibo

Amazon River

Parana River

Lake Titicaca

Tierra del Fuego

Patagonian Steppe

Grand Chaco

Brazilian Highlands

Amazon Basin

Sierra Madre

Orinoco River

Atacama Desert

3. The great agricultural and mineral wealth of South America is not evenly distributed. The export economies of several South American nations are dominated by dependence on one or two commodities. Using your text as a reference and the map on the reverse page, use a seperate color to shade each of the major regions where the following commodities are produced.

oil

copper

tin

bauxite and alumina

coffee

sugar

Miles 0   500   1000

Kilometers 0   500   1000

80° 60° 40°

STUDY GUIDE 10

Mid America

1. Define and note the significance of:

Maya	land reform
Aztec	ejidos
latifundia	haciendas
minifundia	Cuban and Nicaraguan Revolutions
land tenure	micro nation states-economic difficulties

2. Know the location of the following:

Social/Economic/Cultural Features

San Jose
Havana
Kingston
Santo Domingo
Mexico City
Guatemala City
Port-au-Prince
San Salvador
Tegucigalpa
Managua

Physical/Environmental Features

Straits of Florida
Isthmus of Panama
Sierra Madre
Baja California
Yucutan Channel
Gulf of California
Gulf of Mexico
Yucatan
Caribbean Sea
Lesser Antilles
Greater Antilles

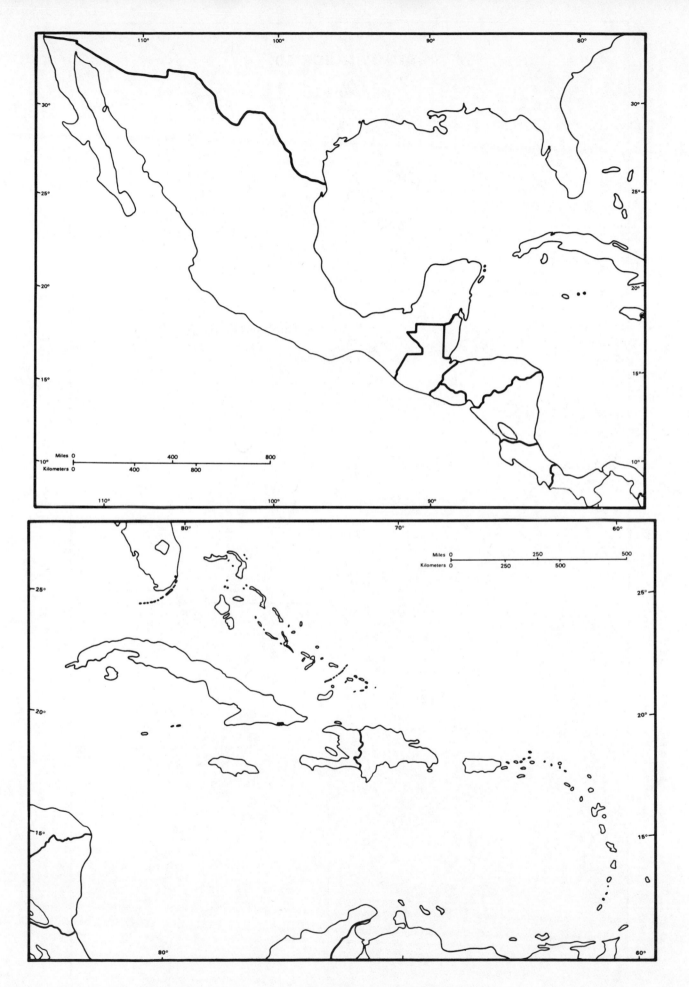

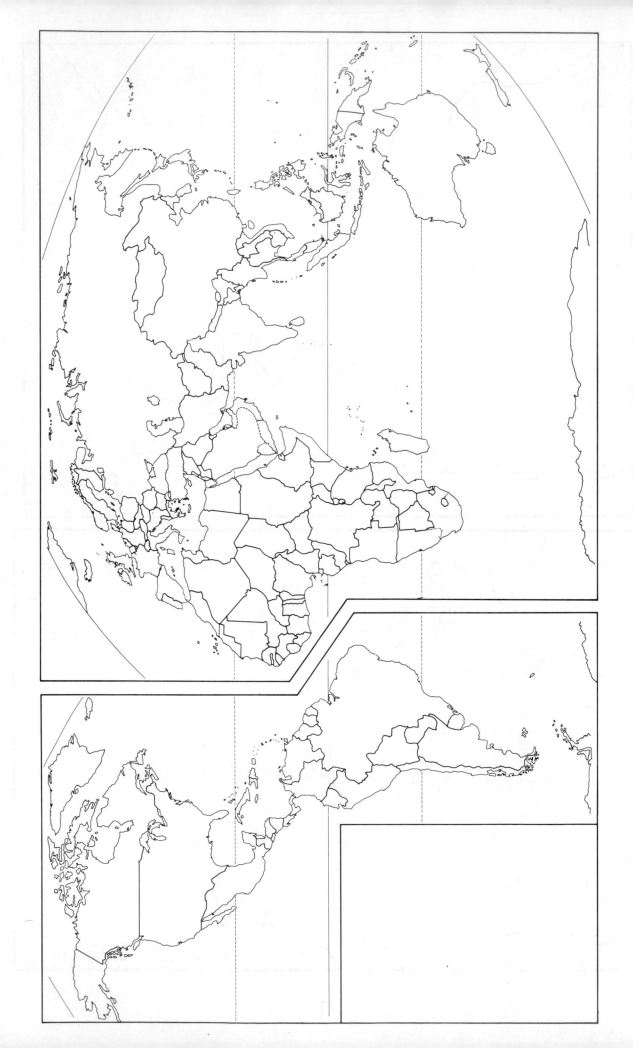

Section 4
Atlas Section

INTRODUCTION

Maps are an essential tool of the geographers. They are often a basic tool for analysis. They certainly are a valuable aid in the presentation of information about, and the description of, places. In most world regional or introductory geography classes students are overwhelmed with maps. The text book invariably makes extensive use of maps; so does the instructor. In the typical regional geography class the instructor, either by sweeping motions of the hand or pointer makes frequent, if not constant, reference to wall maps, chalkboard outlines and/or slide or overhead map images. Students make determined, if unsuccessful efforts to replicate, not only the information in the lecture, but images on the screen, board or wall map. Understandably reluctant to draw or write in their reasonably expensive textbook for fear of reducing its resale value, the student's only recourse is to duplicate both the map and the narration as nearly as possible. How often we have seen students attempting to draft the complicated outline of Europe, Japan or Indonesia. After several years of witnessing this painful process we offer an alternative.

In Section 4, are a series of regional outline maps. These are fairly detailed, black and white outlines which should prove adequate for taking the notes that invariably accompany the detailed classroom discussion of world regions. We encourage you to use these maps in class, to organize your review materials, or just as study aides.

Map 1   North America

74

Map 2 United States

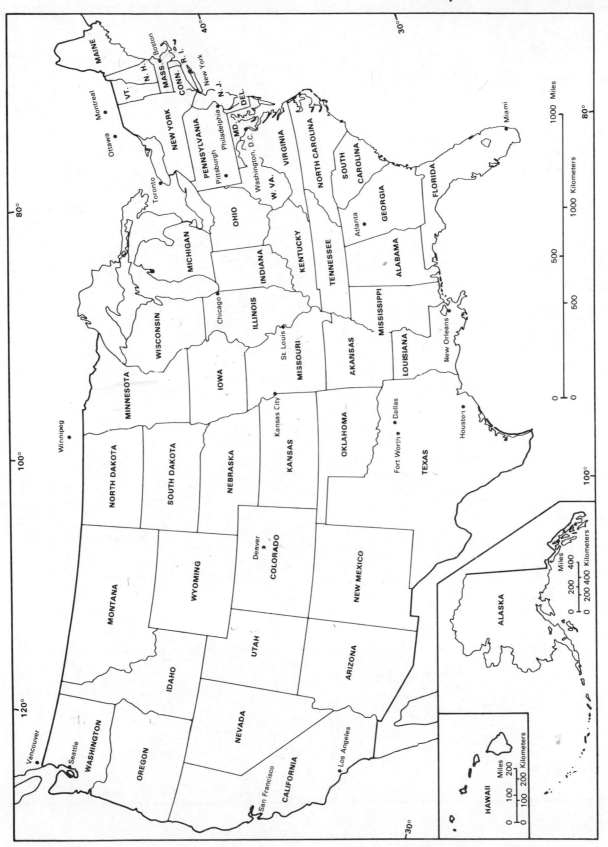

Map 3 Europe

Map 4 USSR

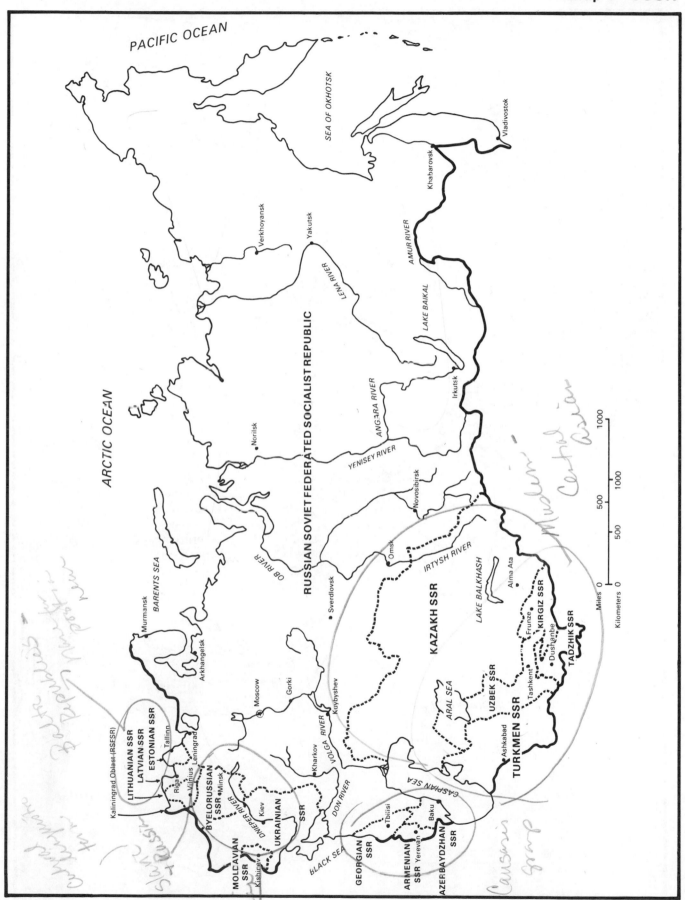

PACIFIC OCEAN

SEA OF OKHOTSK

Vladivostok

Khabarovsk

AMUR RIVER

Verkhoyansk

Yakutsk

LENA RIVER

LAKE BAIKAL

RUSSIAN SOVIET FEDERATED SOCIALIST REPUBLIC

Irkutsk

ANGARA RIVER

ARCTIC OCEAN

Norilsk

YENISEY RIVER

Novosibirsk

OB RIVER

Omsk

IRTYSH RIVER

BARENTS SEA

Sverdlovsk

KAZAKH SSR

Alma Ata

LAKE BALKHASH

Frunze

KIRGIZ SSR

TADZHIK SSR

Murmansk

Dushanbe

Arkhangelsk

Moscow

Gorki

Kuybyshev

ARAL SEA

UZBEK SSR

Tashkent

TURKMEN SSR

Kaliningrad Oblast (RSFSR)

LITHUANIAN SSR

LATVIAN SSR

ESTONIAN SSR

Tallinn

Kharkov

VOLGA RIVER

Ashkabad

Riga

BYELORUSSIAN SSR

Vilnius

Leningrad

Minsk

DNIEPER RIVER

Kiev

UKRAINIAN

SSR

DON RIVER

CASPIAN SEA

Tbilisi

Baku

MOLCAVIAN SSR

Kishinev

GEORGIAN SSR

ARMENIAN SSR

Yerevan

AZERBAYDZHAN SSR

BLACK SEA

Miles 0

500

1000

Kilometers 0

500

1000

77

Map 5 U.S.S.R. and Asia

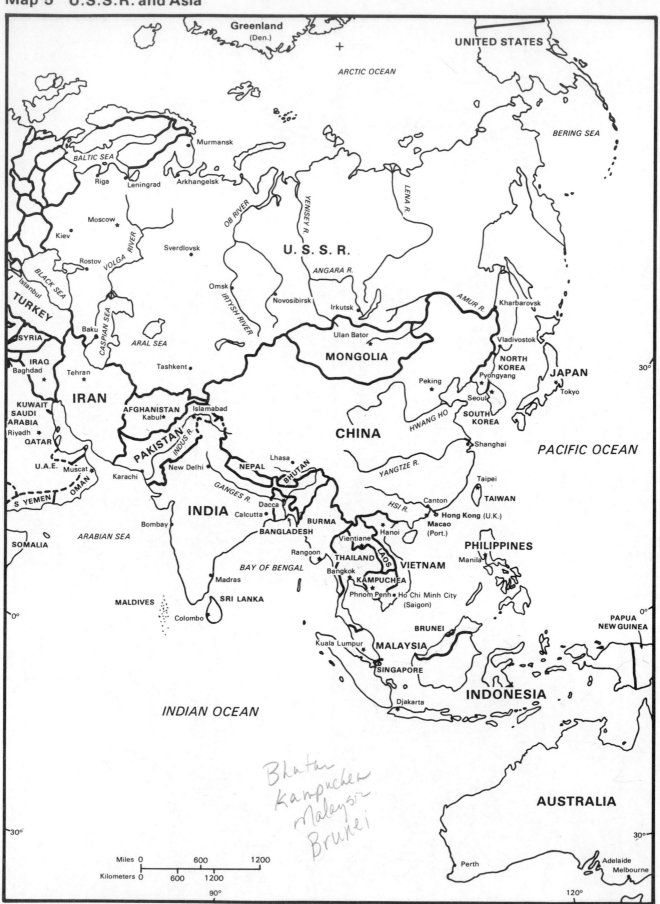

Map 6   China and its Borderlands

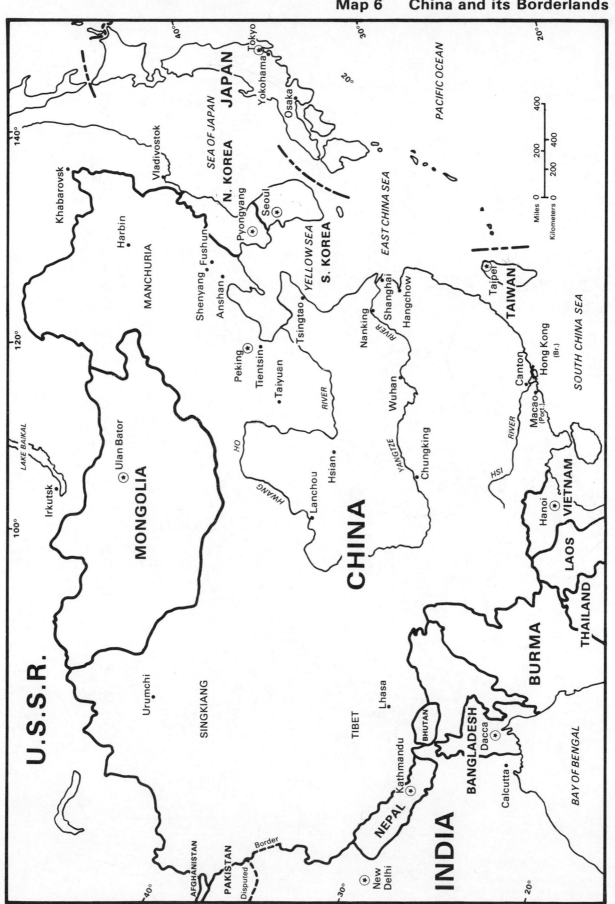

Map 7   Japan

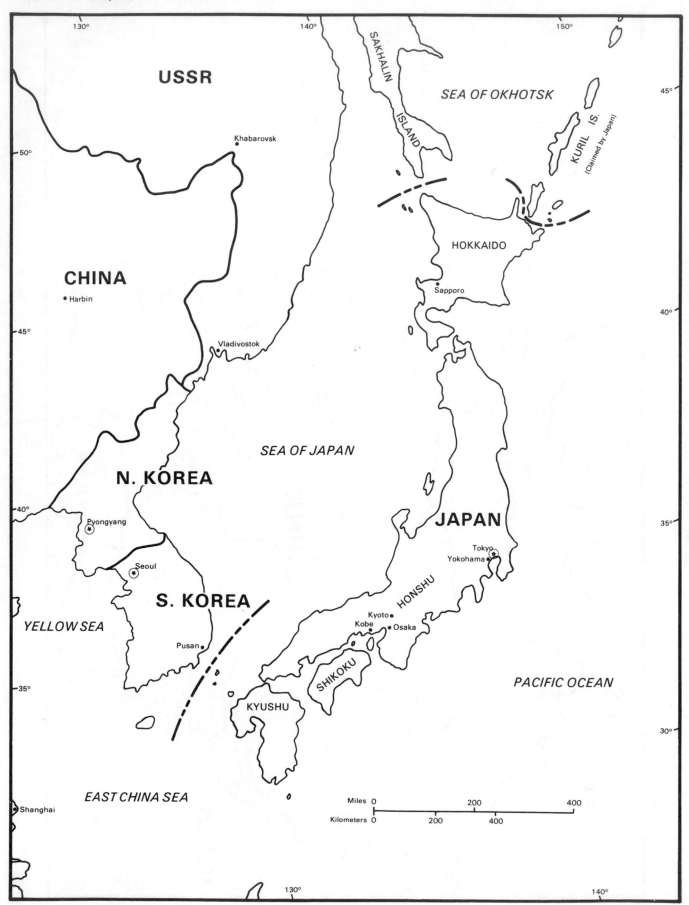

USSR

Khabarovsk

SAKHALIN ISLAND

SEA OF OKHOTSK

KURIL IS.
(Claimed by Japan)

CHINA

Harbin

HOKKAIDO

Sapporo

Vladivostok

SEA OF JAPAN

N. KOREA

Pyongyang

JAPAN

Tokyo
Yokohama

Seoul

S. KOREA

HONSHU

YELLOW SEA

Pusan

Kyoto
Kobe
Osaka

PACIFIC OCEAN

SHIKOKU

KYUSHU

EAST CHINA SEA

Shanghai

| Miles 0 | | 200 | 400 |
| Kilometers 0 | 200 | 400 | |

Map 8   Southeast Asia

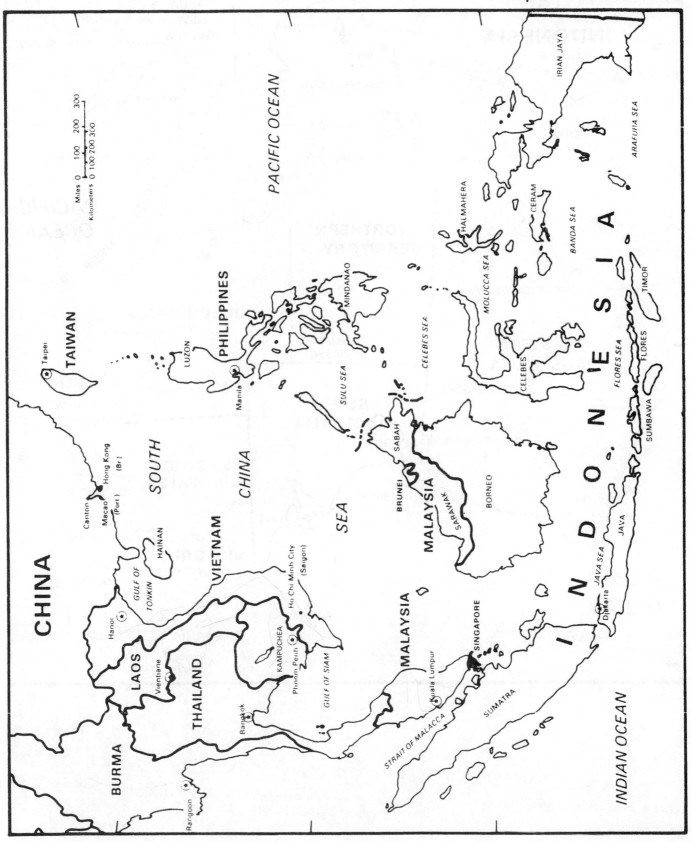

PACIFIC OCEAN

Miles 0   100   200   300

Kilometers 0   100   200   300

IRIAN JAYA

ARAFURA SEA

HALMAHERA

CERAM

BANDA SEA

MOLUCCA SEA

TIMOR

TAIWAN

Taipei

PHILIPPINES

MINDANAO

LUZON

CELEBES SEA

FLORES SEA

CELEBES

FLORES

SUMBAWA

Manila

SULU SEA

I N D O N E S I A

SOUTH

CHINA

SABAH

BORNEO

JAVA

CHINA

Canton

Hong Kong
(Br.)

Macao
(Port.)

HAINAN

VIETNAM

SEA

BRUNEI

MALAYSIA

SARAWAK

JAVA SEA

GULF OF
TONKIN

Ho Chi Minh City
(Saigon)

Djakarta

Hanoi

LAOS

MALAYSIA

SINGAPORE

THAILAND

KAMPUCHEA

Vientiane

Phnom Penh

Kuala Lumpur

Bangkok

GULF OF SIAM

SUMATRA

BURMA

STRAIT OF MALACCA

INDIAN OCEAN

Rangoon

81

Map 9 Australia

BORNEO CELEBES CERAM 130° 140° 150°
120°
BANDA SEA IRIAN JAYA PAPUA NEW GUINEA
INDONESIA NEW BRITAIN
FLORES SEA JAMDENA Lae •
SUMBAWA FLORES ARAFURA SEA
10° TIMOR Port Moresby 10°
TIMOR SEA Darwin CORAL SEA
ARNHEM LAND GULF OF CARPENTARIA PACIFIC OCEAN
NORTHERN TERRITORY GREAT BARRIER REEF
20° GREAT SANDY DESERT 20°
QUEENSLAND
WESTERN AUSTRALIA • Alice Springs Rockhampton •
SIMPSON DESERT
GIBSON DESERT
SOUTH AUSTRALIA Brisbane
GREAT VICTORIA DESERT NEW SOUTH WALES
30° Perth GREAT AUSTRALIAN BIGHT Newcastle • 30°
Fremantle • Sydney
Adelaide Canberra AUST. CAP. TERR.
INDIAN OCEAN VICTORIA TASMAN SEA
Melbourne
Miles 0 250 500 BASS STRAIT
Kilometers 0 250 500 TASMANIA
40° Hobart 40°
120° 140° 160°

Papua New Guinea

82

Map 10   Indian subcontinent

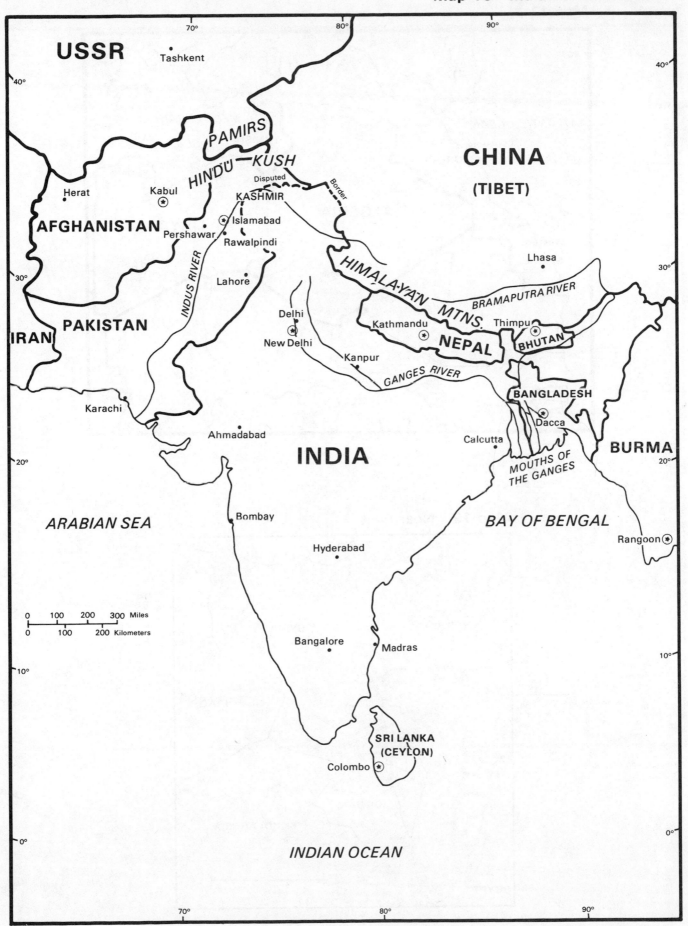

Map 11

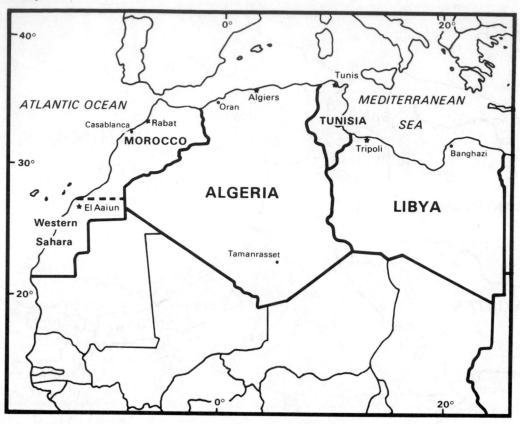

Map 12   Near East

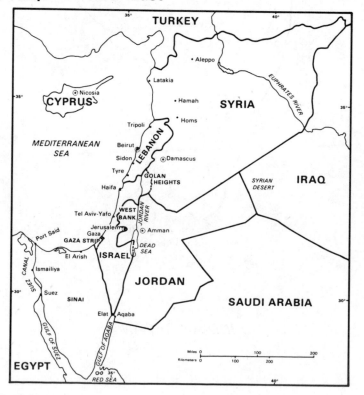

Map 11   The Moslem World

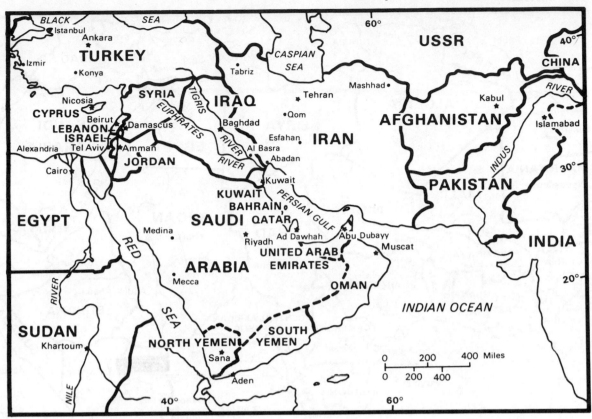

Map 13   Persian Gulf

BLACK SEA
Istanbul
Ankara
TURKEY
Izmir
Konya
60°
CASPIAN SEA
Tabriz
USSR
40°
CHINA
RIVER
Nicosia
SYRIA
CYPRUS
LEBANON
Beirut
Damascus
ISRAEL
Alexandria   Tel Aviv
Amman
JORDAN
Cairo
TIGRIS
EUPHRATES
RIVER
RIVER
IRAQ
Baghdad
Al Basra
Abadan
Kuwait
Tehran
Mashhad
Qom
Esfahan
IRAN
Kabul
AFGHANISTAN
Islamabad
PAKISTAN
INDUS
RIVER
30°

EGYPT
RED
RIVER
SEA
Medina
SAUDI
ARABIA
Mecca
KUWAIT
BAHRAIN
QATAR
Riyadh
Ad Dawhah
UNITED ARAB
EMIRATES
OMAN
PERSIAN GULF
Abu Dubayy
Muscat
INDIAN OCEAN
INDIA
20°

SUDAN
Khartoum
NILE
SEA
NORTH YEMEN
Sana
Aden
SOUTH
YEMEN
0   200   400 Miles
0   200   400
40°
60°

Kermanshah
45°
50°
55°
60°
Baghdad
Karbala
An Najaf
TIGRIS RIVER
EUPHRATES RIVER
IRAQ
Basra
Abadan
KUWAIT
Al Kuwait
Esfahan
IRAN
Yazd
Ahvaz
Kerman
30°
Shiraz
Kharg Is.
30°

SAUDI
ARABIA
PERSIAN GULF
Ad Dammam
BAHRAIN
Al Manamah
QATAR
Ad Dawhah
25°
Riyadh
STRAITS OF
HORMUZ
OMAN
Dubayy
Al Fujayrah
GULF OF OMAN
Abu Dhabi
Muscat
25°
UNITED ARAB EMIRATES
OMAN
Miles 0   150   300
Kilometers 0   150   300
45°
50°
55°

85

Map 14   Africa

Map 15   South America

NICARAGUA 80°   CARIBBEAN SEA   60°   40°
Barranquilla   TRINIDAD & TOBAGO
COSTA   Cartagena   Caracas ⊛
RICA
PANAMA   **VENEZUELA**   *ORINOCO RIVER*   Georgetown ⊛
                                                            Paramaribo
Cali   ⊛ Bogota   **GUYANA**   Cayenne
                                                    SURINAM   **FRENCH**
COLUMBIA   *RIO BRANCO*   **GUIANA**   *NORTH ATLANTIC OCEAN*

EQUADOR ⊛ Quito   *RIO NEGRO*
                                                                            0°
Guayaquil   Manaus   *RIVER*   Belem
       Iquitos   Fortaleza
                             AMAZON
                    RIO *MADIERA* *RIO XINGU*   *RIO*   *TOCANTINS*   *SAO*   *FRANCISCO*
PERU   Recife
                                        BRAZIL
Lima ⊛   Salvador
       Cuzco   *LAKE TITICACA*
              La Paz   Brasilia
SOUTH PACIFIC   ⊛ Sucre **BOLIVIA**   ⊛

                                                     Belo Horizonte
                    PARAGUAY   20°
OCEAN   Antofagasta   *PARANA RIVER*   Vitoria
                    Asuncion ⊛   Sao Paulo
                                                   Rio de Janeiro

              Cordoba   *URUGUAY RIVER*   Porto Alegre
CHILE
Valparaiso   Rosario   **URUGUAY**
Santiago ⊛ Buenos Aires ⊛   ⊛ Montevideo   *SOUTH ATLANTIC OCEAN*
Concepcion **ARGENTINA**   *RIO DE LA PLATA*
       Bahia Blanca

Miles 0   500   1000
Kilometers 0   500   1000

Falkland Islands   40°

South Georgia
(Falkland Is.)

(handwritten: Guyana, Surinam, French Guiana)

Map 16 Mexico

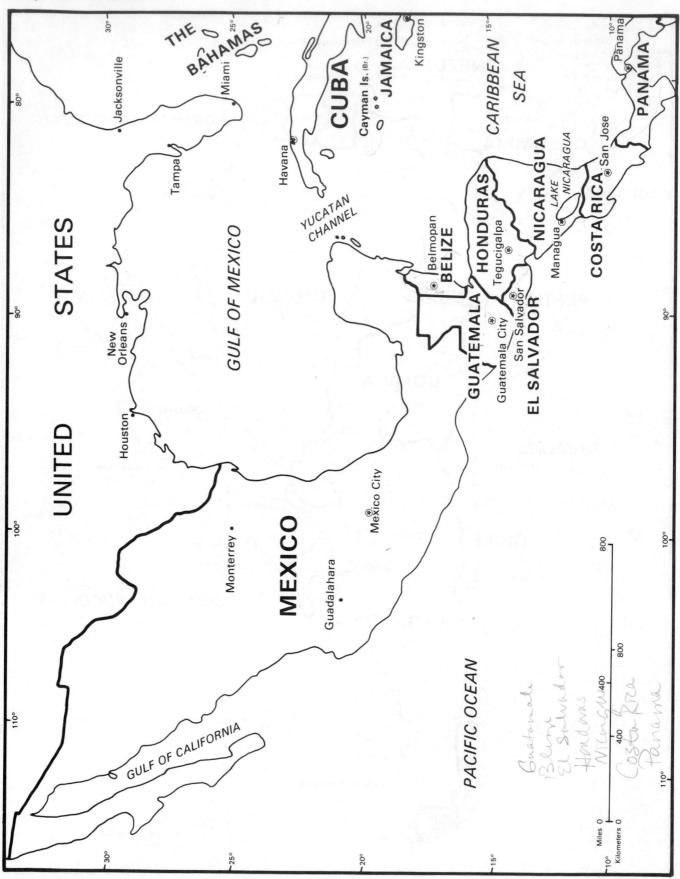

THE BAHAMAS

30° 25° 20° 15° 10°

Jacksonville

80°

Miami

Tampa

Havana

CUBA

Cayman Is. (Br.)

JAMAICA

Kingston

CARIBBEAN SEA

Panama

PANAMA

San Jose

NICARAGUA

LAKE NICARAGUA

COSTA RICA

Managua

HONDURAS

Tegucigalpa

BELIZE

Belmopan

GULF OF MEXICO

YUCATAN CHANNEL

UNITED STATES

90°

New Orleans

Houston

GUATEMALA

Guatemala City

San Salvador

EL SALVADOR

Mexico City

90°

100°

Monterrey

MEXICO

Guadalahara

PACIFIC OCEAN

100°

110°

GULF OF CALIFORNIA

110°

Guatemala
Belize
El Salvador
Honduras
Nicaragua
Costa Rica
Panama

Miles 0
Kilometers 0

400 800
400 800

30° 25° 20° 15° 10°

Map 17   Caribbean

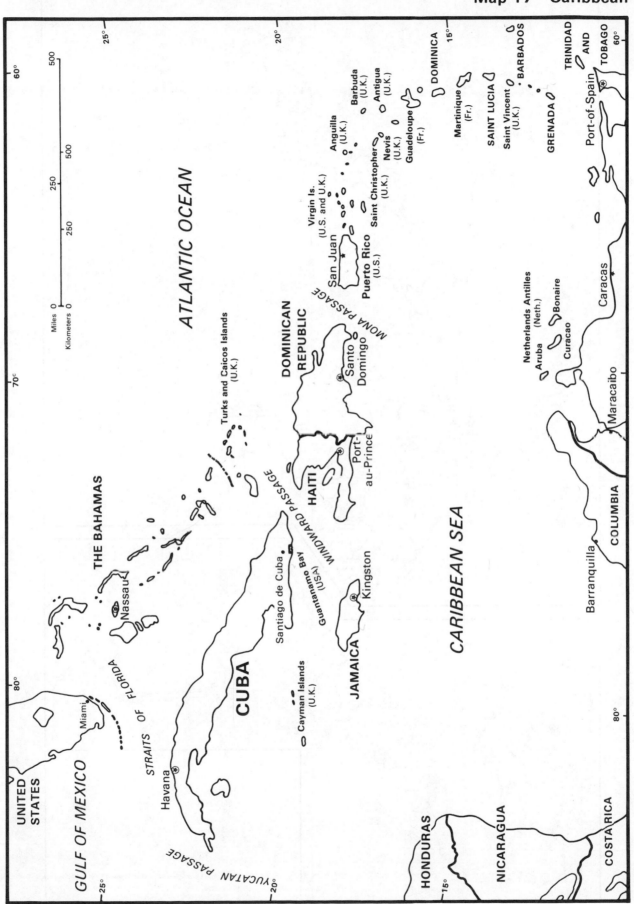

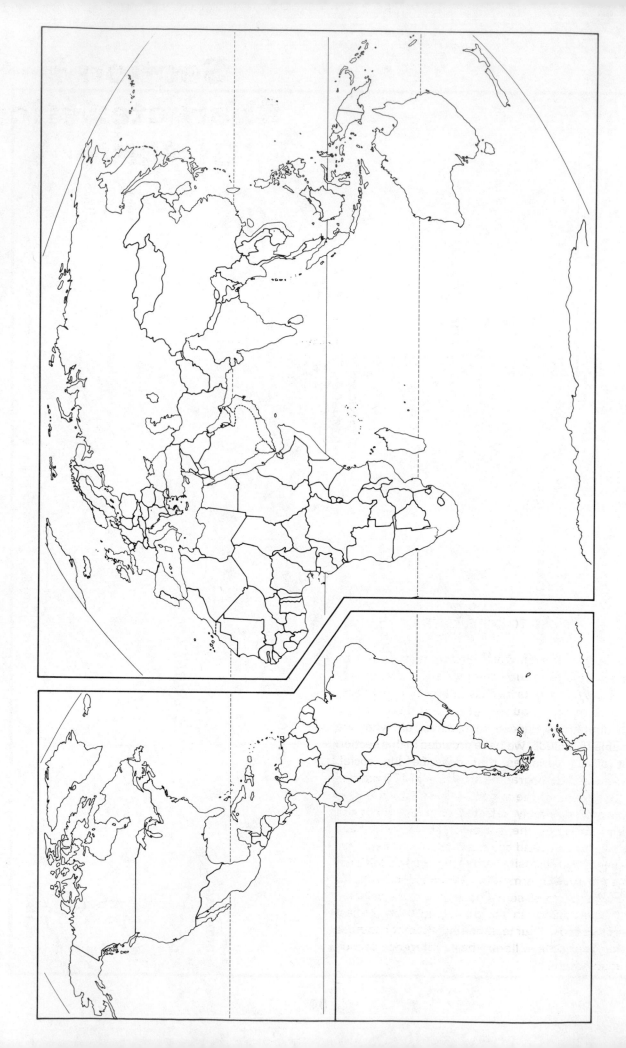

Section 5
Characteristics
Of Nations

A WORD TO THE STUDENT

The exploration of world and regional patterns requires reasonably current information. The quality and availability of data directly influences how finely we are able to tune our images of any place. While space limitations impose restrictions on what we were able to include, we have provided in this section a list of 11 variables that describe the social, economic and demographic structure of 179 nations and 19 regions of the world. The variables in this table were admittedly selected to complement the material on which the exercises of this book have focused. It is not, and could not be, a complete list. However, if you so desire, you may explore additional sources of such information. We hope you will. To encourage you to do so we have provided a selected list of data sources in the bibliography on the last page of this book. Your textbook will list other sources of data. Your college library has a reference section with many more.

Region or Country	1 Population Estimate mid-1985 (millions)	2 Crude Birth Rate	3 Crude Death Rate	4 Natural Increase (Annual Percent)	5 Population "Doubling Time" in Yrs. (at current rate)	6 Per Capita Military Exp. (U.S. $)	7 Infant Mortality Rate	8 Per Capita Arable Land (acres)	9 Urban Population (%)	10 Per Capita Calorie Supply as Percent of Requirements	11 Per Capita GNP, 198 (U.S. $)
WORLD	4845	27	11	1.7	41	—	81	—	41	109	2760
MORE DEVELOPED	1174	15	9	0.6	118	—	18	—	72	134	9380
LESS DEVELOPED	3671	31	11	2.0	34	—	90	—	31	101	710
LESS DEV. (excl. China)	2629	36	12	2.4	29	—	101	—	34	99	890
AFRICA	551	45	16	2.9	24	—	110	—	31	100	750
NORTHERN AFRICA	128	41	12	2.9	24	—	97	—	42	110	1190
Algeria	22.2	45	12	3.3	21		109	.9	52	100	2400
Egypt	48.3	37	10	2.7	26	91.0	80	.2	44	117	700
Libya	4.0	46	11	3.5	20	156.0	92	2.2	64	144	7500
Morocco	24.3	41	12	2.9	24	42.0	99	1.0	42	109	750
Sudan	21.8	46	17	2.9	24	14.0	118	1.1	21	100	400
Tunisia	7.2	33	10	2.3	30	30.0	85	1.3	52	115	1290
WESTERN AFRICA	166	48	18	3.0	23	—	118	—	29	96	580
Benin	4.0	51	23	2.8	25	3.0	149	2.1	39	100	290
Cape Verde	0.3	36	9	2.7	26	—	77	.3	20	—	360
Gambia	0.8	49	29	2.0	35	—	193	1.1	21	94	290
Ghana	14.3	47	15	3.2	22	5.0	107	.2	40	87	320
Guinea	6.1	47	23	2.4	29	4.0	147	2.2	22	83	300
Guinea-Bissau	0.9	41	22	1.9	36	—	143	1.1	27	—	180
Ivory Coast	10.1	46	18	2.9	25	11.0	122	2.6	42	113	720
Liberia	2.2	46	15	3.1	22	5.0	112	.2	39	107	470
Mali	7.7	49	21	2.8	25	4.0	137	3.8	18	84	150
Mauritania	1.9	50	21	2.9	24	30.0	137	.3	35	88	440
Niger	6.5	51	23	2.8	25	2.0	140	7.4	16	94	240
Nigeria	91.2	48	17	3.1	22	28.0	105	.8	28	99	760
Senegal	6.7	50	19	3.1	22	10.0	141	1.1	42	100	440
Sierra Leone	3.6	47	30	1.7	41	2.0	200	3.0	28	91	380
Togo	3.0	45	17	2.8	25	8.0	113	2.3	20	91	290
Upper Volta	6.9	48	22	2.6	27	14.0	149	2.1	8	85	280
EASTERN AFRICA	159	48	17	3.1	23	—	109	—	17	88	300
Burundi	4.6	48	21	2.7	26	5.0	137	.6	7	92	240
Comoros	0.5	46	16	3.0	23	—	88	.6	19	105	—
Djiboudi	0.3	46	20	2.6	27	—	122	—	74	—	—
Ethiopia	36.0	43	22	2.1	33	5.0	142	1.1	15	74	140
Kenya	20.2	54	13	4.1	17	12.0	82	.3	16	88	340
Madagascar	10.0	45	17	2.8	25	6.0	67	.7	22	107	290
Malawi	7.1	52	20	3.2	22	4.0	165	1.0	12	95	210
Mauritius	1.0	21	7	1.5	47	2.0	26.9	.3	43	119	1150
Mozambique	13.9	45	17	2.8	25	9.0	110	.7	13	80	—
Reunion	0.5	23	6	1.7	42	—	14	—	41	—	3710
Rwanda	6.3	53	17	3.6	19	3.0	110	.4	5	94	270
Seychelles	0.1	26	7	1.9	37	—	14.4	—	37	—	2400
Somalia	6.5	47	21	2.6	27	12.0	143	.8	34	92	250
Tanzania	21.7	50	15	3.5	20	10.0	98	.6	14	87	240
Uganda	14.7	50	15	3.5	20	11.0	94	.7	14	79	220
Zambia	6.8	48	15	3.3	21	41.0	101	2.3	43	86	580
Zimbabwe	8.6	47	12	3.5	20	31.0	70	.9	24	79	740
MIDDLE AFRICA	62	45	18	2.7	26	—	119	—	34	95	420
Angola	7.9	47	22	2.5	28	—	149	.5	24	89	—
Cameroon	9.7	44	18	2.6	27	7.0	117	2.1	42	105	800
Central African Republic	2.7	46	22	2.4	29	4.0	143	5.5	41	95	280
Chad	5.2	44	23	2.1	33	8.0	143	4.0	22	75	—
Congo	1.7	44	19	2.5	28	23.0	124	1.1	48	99	1230
Equatorial Guinea	0.3	43	21	2.2	32	25.0	137	.9	60	—	—
Gabon	1.0	35	18	1.7	41	78.0	112	1.2	41	121	4250
Sao Tome & Principe	0.1	39	10	2.9	24	—	69.2	.03	32	—	310
Zaire	33.1	45	16	2.9	24	10.0	106	.5	34	96	160
SOUTHERN AFRICA	37	36	14	2.2	31	—	94	—	52	115	2280
Botswana	1.1	50	13	3.7	19	22.0	79	4.6	16	—	920
Lesotho	1.5	42	16	2.6	27	—	110	.7	6	—	470
Namibia	1.1	45	17	2.8	25	—	115	—	51	—	1760
South Africa	32.5	35	14	2.1	33	80.0	92	1.3	56	115	2450
Swaziland	0.6	48	17	3.1	22	2.0	129	.7	26	—	890

Region or Country	1 Population Estimate mid-1985 (millions)	2 Crude Birth Rate	3 Crude Death Rate	4 Natural Increase (Annual Percent)	5 Population "Doubling Time" in Yrs. (at Current rate)	6 Per Capita Military Exp. (U.S. $)	7 Infant Mortality Rate	8 Per Capita Arable Land (acres)	9 Urban Population (%)	10 Per Capita Calorie Supply as Percent of Requirements	11 Per Capita GNP, 1983 (U.S. $)
ASIA	2829	28	10	1.8	39	—	87	—	27	100	950
SOUTHWEST ASIA	114	39	11	2.8	25	—	94	—	53	113	3500
Bahrain	0.4	32	5	2.7	26	141.0	37	.007	81	—	10,360
Cyprus	0.7	21	9	1.2	57	39.0	17	1.5	53	128	3720
Gaza	0.5	46	8	3.8	18	—	109	—	90	—	—
Iraq	15.5	46	13	3.3	21	159.0	72	1.0	68	109	—
Israel	4.2	24	6	1.8	39	839.0	14.2	.2	87	118	5360
Jordan	3.6	46	8	3.8	18	87.0	63	1.0	60	97	1710
Kuwait	1.9	35	3	3.2	22	613.0	22.8	.002	90	—	18,180
Lebanon	2.6	29	9	2.1	34	58.0	48	.2	76	100	—
Oman	1.2	47	16	3.1	22	914.0	122	.05	7	—	6240
Qatar	0.3	31	3	2.8	25	1194.0	45	.02	86	—	21,170
Saudi Arabia	11.2	42	12	3.0	23	1004.0	103	.3	70	119	12,180
Syria	10.6	47	7	3.9	18	147.0	57	1.6	47	113	1680
Turkey	52.1	35	10	2.5	28	64.0	110	1.4	45	117	1230
United Arab Emirates	1.3	27	4	2.3	30	836.0	45	.02	81	117	21,340
Yemen, North	6.1	48	21	2.7	26	52.0	154	.7	15	93	510
Yemen, South	2.1	48	19	2.9	24	43.0	138	.3	37	87	510
MIDDLE SOUTH ASIA	1058	37	13	2.3	30	—	120	—	24	91	260
Afghanistan	14.7	48	23	2.5	28	5.0	205	1.3	16	75	—
Bangladesh	101.5	45	17	2.8	25	1.0	133	.3	15	84	130
Bhutan	1.4	38	18	2.0	35	—	144	.5	5	—	—
India	762.2	34	13	2.2	32	5.0	118	.6	23	90	260
Iran	45.1	41	10	3.0	23	261.0	101	1.1	50	120	—
Maldives	0.2	43	13	3.0	23	—	108	—	20	—	—
Nepal	17.0	42	18	2.4	29	1.0	144	.4	6	86	170
Pakistan	99.2	43	15	2.7	25	12.0	120	.6	29	99	390
Sri Lanka	16.4	27	6	2.1	33	1.0	34.4	.2	22	101	330
SOUTHEAST ASIA	404	32	10	2.2	31	—	79	—	24	103	710
Brunei	0.2	28	4	2.4	29	642.0	15	—	64	—	21,140
Burma	36.9	37	15	2.2	32	5.0	94	.7	24	105	180
Democratic Kampuchea	6.2	32	11	2.1	33	—	160	.8	16	80	—
East Timor	0.7	48	23	2.5	28	—	183	—	12	—	—
Indonesia	168.4	34	12	2.2	32	11.0	87	.3	22	106	560
Laos	3.8	41	18	2.3	30	10.0	122	.7	16	83	—
Malaysia	15.7	29	7	2.2	32	45.0	29	.6	32	118	1870
Philippines	56.8	32	7	2.5	28	11.0	50	.3	37	102	760
Singapore	2.6	16	5	1.1	64	186.0	9.4	.002	100	—	6620
Thailand	52.7	25	6	1.9	36	18.0	51	.9	17	103	810
Vietnam	60.5	34	9	2.5	28	18.0	90	.2	19	93	—
EAST ASIA	1252	18	8	1.1	65	—	35	—	29	106	1380
China	1042	19	8	1.1	65	26.0	38	.3	21	105	290
Hong Kong	5.5	15	5	1.0	67	—	9.9	—	92	—	6000
Japan	120.8	13	6	0.6	110	80.0	6.2	.09	76	123	10,100
Korea, North	20.1	31	7	2.3	30	63.0	32	.3	64	127	—
Korea, South	42.7	23	6	1.7	41	75.0	29	.1	57	124	2010
Macao	0.3	25	7	1.8	38	—	38	—	97	—	2560
Mongolia	1.9	34	7	2.7	26	76.0	50	1.7	51	—	—
Taiwan	19.2	21	5	1.6	44	106.0	8.9	.2	71	—	—
NORTH AMERICA	264	15	8	0.7	99	—	10	—	74	137	13,890
Canada	25.4	15	7	0.8	90	174.0	9.1	4.6	76	126	12,000
United States	238.9	16	9	0.7	100	499.0	10.5	2.1	74	138	14,090
LATIN AMERICA	406	31	8	2.3	30	—	62	—	66	108	1890
MIDDLE AMERICA	105	33	6	2.7	26	—	55	—	63	115	1940
Belize	0.2	32	7	2.5	28	—	27	—	50	—	1140
Costa Rica	2.6	31	4	2.7	26	11.0	19.3	.3	48	117	1020
El Salvador	5.1	28	6	2.1	33	13.0	42.2	.3	39	94	710
Guatemala	8.0	43	8	3.5	20	9.0	62.4	.5	39	94	1120
Honduras	4.4	44	10	3.4	20	11.0	82	.5	37	96	670
Mexico	79.7	32	6	2.6	27	8.0	53	.8	70	120	2240
Nicaragua	3.0	44	10	3.4	20	28.0	76	1.4	53	101	900
Panama	2.0	25	5	2.0	34	9.0	26	.6	49	99	2070
CARIBBEAN	31	25	8	1.8	39	—	55	—	54	103	—
Antigua and Barbuda	0.1	15	5	1.0	67	—	11.1	—	34	—	1730
Bahamas	0.2	24	5	1.9	37	—	24.7	.02	65	—	4060
Barbados	0.3	18	8	1.0	72	4.0	14.2	.3	42	—	3930
Cuba	10.1	17	6	1.1	64	49.0	16.8	.6	70	117	—
Dominica	0.1	22	5	1.7	40	—	12.6	.2	—	—	970
Dominican Republic	6.2	33	8	2.5	28	17.0	64	.4	52	94	1380
Grenada	0.1	25	7	1.8	39	—	15.4	.05	—	—	990
Guadeloupe	0.3	20	7	1.4	51	—	23	—	46	—	—
Haiti	5.8	36	13	2.3	30	2.0	108	.3	28	83	320
Jamaica	2.3	28	6	2.2	32	9.0	28	.2	54	114	1300
Martinique	0.3	18	7	1.1	62	—	20	—	71	—	4270
Netherlands Antilles	0.3	19	6	1.3	53	—	26	—	66	—	—
Puerto Rico	3.3	20	6	1.3	52	—	16.0	—	67	—	2890
St. Lucia	0.1	31	6	2.5	28	—	26.1	.1	40	—	1060
St. Vincent & the Grenadines	0.1	26	6	2.0	35	—	46.8	.3	—	—	860
Trinidad & Tobago	1.2	25	7	1.9	37	11.0	28	.1	23	111	6900

Region or Country	1 Population Estimate mid-1985 (millions)	2 Crude Birth Rate	3 Crude Death Rate	4 Natural Increase (Annual Percent)	5 Population "Doubling Time" in Yrs. (at current rate)	6 Per Capita Military Exp. (U.S. $)	7 Infant Mortality Rate	8 Per Capita Arable Land (acres)	9 Urban Population (%)	10 Per Capita Calorie Supply as Percent of Requirements	11 Per Cap GNP, 19 (U.S. $
TROPICAL SOUTH AMERICA	225	32	8	2.4	29	—	70	—	66	103	1860
Bolivia	6.2	42	16	2.7	26	18.0	124	1.5	46	87	510
Brazil	138.4	31	8	2.3	30	18.0	71	.7	68	105	1890
Colombia	29.4	28	7	2.1	33	7.0	53	.4	67	106	1410
Ecuador	8.9	35	8	2.7	26	22.0	70	1.3	45	91	1430
Guyana	0.8	28	6	2.2	31	10.0	35	1.1	32	109	520
Paraguay	3.6	35	7	2.8	25	13.0	45	.8	39	125	1410
Peru	19.5	35	10	2.5	28	33.0	99	.5	65	92	1040
Suriname	0.4	28	8	2.0	34	—	31	.2	66	109	3520
Venezuela	17.3	33	6	2.7	25	44.0	39	.9	76	107	4100
TEMPERATE SO. AMERICA	46	24	8	1.6	44	—	32	—	83	122	2020
Argentina	30.6	24	8	1.6	44	55.0	35.3	2.3	83	127	2030
Chile	12.0	24	6	1.8	39	73.0	23.6	1.3	83	112	1870
Uruguay	3.0	18	9	0.9	77	40.8	33.2	1.6	84	107	2490
EUROPE	492	13	10	0.3	240	—	15	—	73	135	8200
NORTHERN EUROPE	83	13	11	0.1	465	—	9	—	75	130	9680
Denmark*	5.1	10	11	-0.1	—	259.0	8.2	1.2	83	130	11,490
Finland	4.9	14	9	0.5	154	102.0	6.0	1.3	60	—	10,440
Iceland	0.2	19	7	1.2	60	—	7.1	.09	89	—	10,270
Ireland*	3.6	19	9	1.0	71	59.0	10.5	.8	56	149	4310
Norway	4.2	12	10	0.2	365	322.0	7.8	.5	71	122	13,820
Sweden	8.3	11	11	0.0	6930	365.0	7.0	.9	83	117	12,400
United Kingdom*	56.4	13	12	0.1	630	262.0	10.1	.3	76	131	9050
WESTERN EUROPE	155	12	11	0.1	729	—	10	—	83	134	10,870
Austria	7.5	12	12	0.0	—	93.0	12.0	.5	55	132	9210
Belgium*	9.9	12	11	0.1	1155	322.0	11.3	.2	95	149	9160
France*	55.0	14	10	0.4	198	350.0	9.0	.8	73	134	10,390
Germany, West*	61.0	10	11	-0.2	—	350.0	10.1	.3	94	132	11,420
Luxembourg*	0.4	12	11	0.0	3465	103.0	11.2	.2	78	149	12,190
Netherlands*	14.5	12	8	0.4	192	304.0	8.4	.1	88	129	9910
Switzerland	6.5	11	9	0.2	330	280.0	7.7	.1	58	131	16,390
EASTERN EUROPE	112	16	11	0.5	140	—	19	—	62	136	—
Bulgaria	8.9	14	11	0.2	315	77.0	16.8	1.1	65	145	—
Czechoslovakia	15.5	15	12	0.3	248	143.0	15.6	.8	74	140	—
Germany, East	16.7	14	13	0.1	990	218.0	10.7	.7	77	142	—
Hungary	10.7	12	14	-0.2	—	79.0	19.0	1.2	54	134	2150
Poland	37.3	20	10	1.0	68	99.0	19.3	1.0	59	135	—
Romania	22.8	15	10	0.5	131	56.0	28.0	1.1	49	128	—
SOUTHERN EUROPE	143	13	9	0.4	165	—	18	—	69	139	4810
Albania	3.0	28	6	2.2	32	60.0	43	.5	34	117	—
Greece*	10.1	14	9	0.5	154	220.0	14.9	.8	70	145	3970
Italy*	57.4	11	10	0.1	990	112.0	12.4	.4	72	144	6350
Malta	0.4	15	8	0.7	103	24.0	14.1	.1	85	122	3710
Portugal	10.3	14	9	0.5	136	64.0	19.8	.8	30	130	2190
Spain	38.5	13	7	0.6	116	67.0	9.6	1.1	91	135	4800
Yugoslavia	23.1	17	10	0.7	99	105.0	31.7	.8	46	138	2570
USSR	278	20	10	1.0	71	394.0	32	2.1	64	132	6350
OCEANIA	24	21	8	1.2	56	—	39	—	71	122	8570
Australia	15.8	16	7	0.9	82	207.0	10.3	7.8	86	120	10,780
Fiji	0.7	31	6	2.5	28	12.0	28.8	.6	37	—	1790
French Polynesia	0.2	32	6	2.6	27	—	23	—	57	—	8190
New Zealand	3.3	16	8	0.8	90	93.0	12.5	.3	83	132	7410
Papua-New Guinea	3.3	40	14	2.7	26	9.0	98	.01	13	—	790
Samoa, Western	0.2	34	8	2.7	26	—	49	.9	22	—	—
Solomon Islands	0.3	43	12	3.2	22	—	86	.6	9	—	640
Vanuatu	0.1	40	13	2.8	25	—	98	—	18	—	—

*The ten countries of the common market.

Source: The data in columns 1-5, 7, 9, and 11, is from *The World Population Data Sheet,* The Population Reference Bureau, Inc., Washington, D.C., 1985. The data in columns 6 and 8 is from *The World Almanac and Book of Facts,* Newspaper Enterprise Association, Inc., New York, 1983.

LIST OF SELECTED REFERENCES

Brown, Larry and Gail Finterbusch. *Man, Food and Environment*. as cited in William W. Murdoch *Environment Resources, Pollution and Society*. Stamford Connecticut: Sinaver Associates Inc., 1971.

Brown, Lester R., *The Twenty-Ninth Day*. New York: W.W. Norton and Company, Inc., 1978.

Brown, Lester R., "Population Policies for a New Economic Era." *Worldwatch Paper 53*. Worldwatch Institute, 1983.

Coale, Ansley J., "The History of Human Population." *Scientific American*. Volume 235. Number 3. September, 1974. pp. 40-51.

Freedman, Ronald and Bernard Berelson. "The Human Population." *Scientific American*. Volume 235. Number 3. September, 1974. pp. 30-40.

Murdoch, William W., *Environment Resources, Pollution and Society.* Stamford Connecticut: Sinaver Associates, Inc., 1971.

Rostow, W.W., *The Stages of Economic Growth*. Cambridge: At the University Press, 1971.

Thompson, Warren. "Population." *American Journal of Sociology*. Volume 34. Number 6, 1929. pp. 959-975. as cited in John R. Weeks *Population An Introduction to Concepts and Issues*. Belmont, California: Wadsworth Publishing Company, 1981.

Weeks, John R. *Population An Introduction to Concepts and Issues*. Belmont, California: Wadsworth Publishing Company, 1981.

SOURCES OF DATA

Information Please Almanac, Atlas and Yearbook, 1985, 38th Edition, Simon and Schuster, New York.

U.S. Department of Agriculture, Economic Research Service, *World Agriculture: Outlook and Situation*. Washington, D.C.: U.S. Government Printing Office, March 1983.

The World Almanac and Book of Facts, 1985, Newspaper Enterprise Association, Inc. New York.

World Development Report 1983. The World Bank. London: Oxford University Press.

1985 World Population Data Sheet of the Population Reference Bureau, Inc., Washington, D.C., *1985*.

World Tables, 2nd Edition; 1980, from the data files of the World Bank, Johns Hopkins University Press, Baltimore.

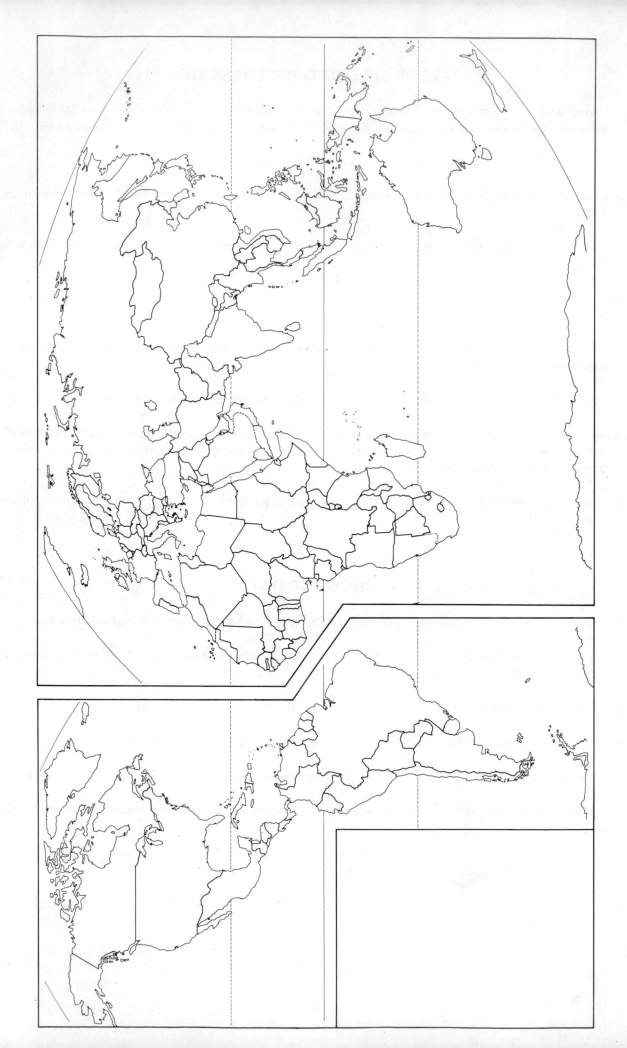